SpringerBriefs in Agriculture

SpringerBriefs in Agriculture present concise summaries of cutting-edge research and practical applications across a wide spectrum of topics in agriculture with a fast turnaround time to publication. Featuring compact volumes of 50 to 125 pages, the series covers a range of content from professional to academic. Monographs of new material are considered for the SpringerBriefs in Agriculture series. Typical topics might include: A timely report of state-of-the art analytical techniques, a bridge between new research results, as published in journal articles, and a contextual literature review, a snapshot of a hot or emerging topic, an in-depth case study or technical example, a presentation of core concepts that students must understand in order to make independent contributions. Best practices or protocols to be followed. A series of short case studies/debates highlighting a specific angle.

Muhammad Azhar Iqbal

Digital Agriculture

An Introduction

 Springer

Muhammad Azhar Iqbal
University of Leeds
Leeds, UK

ISSN 2211-808X   ISSN 2211-8098 (electronic)
SpringerBriefs in Agriculture
ISBN 978-3-031-67678-9   ISBN 978-3-031-67679-6 (eBook)
https://doi.org/10.1007/978-3-031-67679-6

With profound gratitude, to all individuals, from diligent farmers to cherished family members, who steadfastly engage in the noble tasks of cultivating, packaging, supplying, and presenting food, from the bountiful fields to the dining tables.

Preface

Through discussions with academicians and undergraduate students rooted in agricultural disciplines, it is observed that instructors and students alike (particularly in developing nations) have little knowledge about the availability and utilization of Information and Communication Technologies (ICT) in agriculture. Therefore, I discerned a dire need to introduce them the fundamental principles, underlying concepts, and benefits regarding the availability and utilization of advanced digital technologies and techniques that are practiced in developed countries and ensure environmental sustainability with higher agricultural yields at lower cost. This book aims to provide a foundational resource for instructors and undergraduate students devoid of prior exposure to digital technologies in agriculture. The key feature of this book is its elucidation of the significance and utilization of digital technologies and methodologies in managing the flow of agricultural information from the data acquisition phase to the data analysis phase while offering a user-friendly approach to understanding the design of modern digital agriculture systems. Each chapter concludes with some case study questions designed to assess readers' comprehension of the discussed concepts in that chapter. To the best of my knowledge, this book marks the first endeavor to explain the impact of advanced technologies on the digitalization of agricultural operations while considering all phases of an ICT-based digital agriculture ecosystem. Therefore, it can be considered a foundational textbook at the undergraduate level, especially in universities where dedicated modules on agricultural digitalization are still not part of the curriculum. To address the issues related to the understanding of digital agriculture fundamentals, this book is structured as follows:

Chapter 1 serves as an introduction while delineating the evolution, revolutions, driving factors, vision, definition, types, and framework of digital agriculture systems. Chapter 2 delves into the placement of enabling digital technologies essential for automating the monitoring and management of agricultural operations with a focus on the fundamental aspects of agricultural data digitalization across various phases of agricultural production. The discussion extends to the layered architecture of IoT-based Agricultural Ecosystems, laying the groundwork for subsequent chapters.

Chapter 3 revolves around data acquisition in agricultural settings, explicating the roles of various digital acquisition technologies such as RFID (Radio Frequency Identification), sensors, RFID-based sensors, smartphones, remote sensing platforms, and development boards.

Chapter 4 explores the role of short-range, long-range, cellular, satellite, and broadband communication technologies in transmitting data from agricultural fields to cloud servers.

Chapter 5 sheds light on the concept of agricultural Big Data, introducing key enabling technologies for its storage and processing on edge, fog, and cloud servers.

Chapter 6 outlines basic analytical techniques, both statistical and machine learning-based, and visualization technologies crucial for analyzing agricultural Big Data.

Chapter 7 underscores the benefits of Digital Agriculture and delineates future directions in this domain.

Leeds, UK   Muhammad Azhar Iqbal

Contents

About the Author

Muhammad Azhar Iqbal completed his Ph.D. in Communication and Information Systems in 2012 from Huazhong University of Science and Technology (HUST), China. Later, he worked in different universities on different positions. Prior to his current tenure as Assistant Professor at the University of Leeds, he held significant roles (as Teaching Fellow) at Lancaster University (LU, United Kingdom), (as Associate Professor) at Southwest Jiaotong University (SWJTU, China), and (as Associate Professor) at Capital University of Science and Technology (CUST, Pakistan). He has received senior membership IEEE and fellowship of the Higher Education Academy. He has authored several international conferences/journal publications and leading author of two books on the topic of Network Simulations and Internet of Things (IoT). His current research interest is at the cutting edge of agriculture digitalization, and he is dedicated to developing artificial intelligence-based solutions aimed at improving sustainable crop/animal production systems. e-mail: m.a.iqbal1@leeds.ac.uk

Chapter 1
Fundamentals of Digital Agriculture

1.1 Learning Objectives

After studying this chapter, students will be able to

- articulate the definition of Agriculture
- describe the brief history and evolution of agricultural farming systems
- indicate the correspondence of agricultural revolutions with industrial revolutions
- explain the concept of Digital Agriculture
- elaborate the types of Digital Agriculture
- point out the components of the Digital Agriculture Framework

1.2 Agriculture

Agriculture serves the fundamental demands and necessities of humans with the provisioning of food (i.e., crops, vegetables, fruits, herbs, spices, beverages, plants, meat, milk, eggs, honey, oil, etc.), clothing (i.e., cotton, wool, silk, leather, etc.), and sheltering (i.e., lumber, carpeting, plastics, etc.). Likewise, it helps humans with medicine, household items, fuel, and recreation. The word Agriculture is derived from the Latin word *Agricultura*, which is ultimately a combination of two Latin words *Ager (means land or field)* and *Cultura (means cultivation or growing)*. Thus, in the literal sense, the word Agriculture means the cultivation of land. However, the agriculture study is not only concerned with crop cultivation and planting trees but also with rearing livestock (Harris and Fuller 2014; Chandrasekaran et al. 2010). Consequently, Agriculture can be defined as "*the art and science of cultivating soil, growing crops, and rearing livestock for economic purposes*" (Oxford lexico dictionary. https:// www.lexico.com/en/definition/agriculture; National geographic society. https://edu cation.nationalgeographic.org/resource/the-art-and-science-of-agriculture/). As an *Art*, agriculture emphasizes the utilization of the knowledge and skills to perform

M. A. Iqbal, *Digital Agriculture*,
SpringerBriefs in Agriculture, https://doi.org/10.1007/978-3-031-67679-6_1

cultivation and farming of crops and livestock to attain optimum yield. From this perspective, it includes the applicability of prior knowledge with physical (i.e., sowing, plowing, spraying of pesticide and fertilizer, etc.) and mental (i.e., crop selection, plowing method, sowing technique, spatial and temporal aspects of cropping system, etc.) skills. On the other hand, as a *Science*, agriculture promises to employ advanced technologies (based on scientific principles) for the production, management, and protection of crops/plants/trees and livestock to obtain optimum yield while considering the climatic and geographical conditions of a certain area. From a scientific perspective, it includes the adoption of modern techniques for all essential agricultural practices i.e., land management, crop cultivation, crop production, crop protection, fertilization, breeding, hybridization, etc. In recent times, these modern techniques rely heavily on digital and computing technologies e.g. sensing technology, wireless communication, and broadband (Internet) technology for data collection and transmission, fog and cloud computing models for systematic storage of collected data, use of artificial intelligence to improve data analysis efficiency and accuracy, etc. This trend of agricultural digitalization brings innovations to all branches of agriculture (mainly classified into five categories i.e., Crop/Plant/Tree Production and Management, Crop/Plant/Tree Improvement, Crop/Plant/Tree Protection, Animal/Livestock Farming, Allied Disciplines (Chandrasekaran et al. 2010) as shown in Fig. 1.1) and helps agriculturists to work more precisely, efficiently, and sustainably. Before delving into the details of all the innovations determining the current era of "agricultural digitalization", it is important to discuss the evolutionary process of agriculture that supports the adoption of different tools, techniques, and technologies to fulfill the domestic and global demands of food production in different eras of time.

1.3   Evolution of Agricultural Systems/Farming

Since its inception, agriculture has been an evolutionary process and people adopted different approaches to cultivating land and livestock rearing. Below, the history of different farming systems and agricultural revolutions provides a brief overview to help you understand the patterns and characteristics of different farming systems.

1.3.1   Brief History of Farming Systems

Around 12,000 years ago, mankind started agriculture with the collection of grains and the hunting of wild animals for food. Later, human beings took the initiative to cultivate crops and domestication of animals to satisfy the demand for growing families. Since then, a plethora of organized and systematic ways have been adopted to increase the agricultural yield that was ultimately required to satisfy the food demands of the growing human population (Chandrasekaran et al. 2010; Thrall et al.

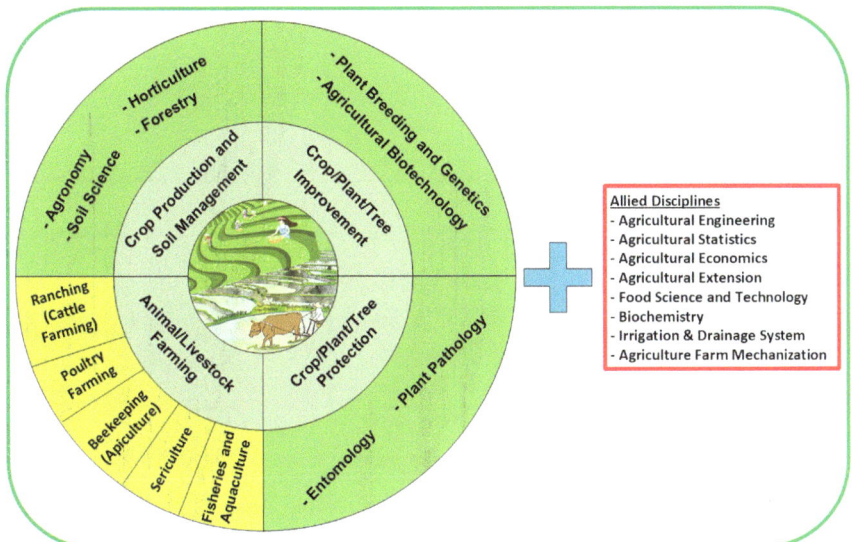

Fig. 1.1 Classification of the branches of agriculture

2010). At an abstract level, it can be stated that *the systematic way of growing crops and raising livestock for business is known as the agricultural system or Farming.* Initially, farming was started individualistically in different regions of the world, but later different ways of farming have been adopted by humans to fulfill their demands for food, clothing, and shelter. Major aspects of different agricultural systems or farming (shown in Fig. 1.2) have been discussed below.

1.3.1.1 Nomadic Farming

Nomadic farming is concerned with the collection and eating of wild grains and wild animals without domestication. Nomads having no fixed habitation considered seasonal changes to move from one place to another and performed cultivation and farming of crops and livestock near forests or rivers.

1.3.1.2 Shifting Cultivation or Assartage Agricultural System

With civilization, humans started the cultivation of founder crops (i.e., wheat, barley, peas, lentils, etc.) and laid down the foundation of shifting cultivation (also known as Swidden Agriculture or Assartage System). The shifting cultivation way of farming deals with the cultivation of a crop in a certain area until the land of that area is worn out. Therefore, a piece of land is cultivated to grow crops for a short period of time (until the soil shows exhaustion signs) and then abandoned to revert to its

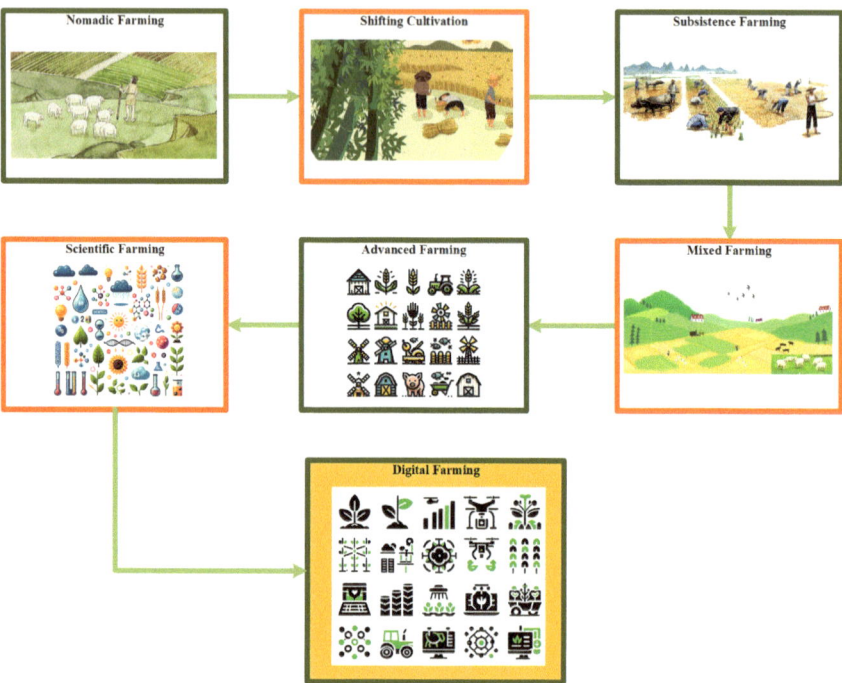

Fig. 1.2 Types of agricultural (farming) systems

natural vegetation. Later, the farmer might have the option to come back to reutilize the fertility of former lands. It is an unsustainable way of farming and scarcity of fertile land is one of the main reasons to discourage its practice in modern times.

1.3.1.3 Subsistence Farming

It is a type of primitive agriculture where farming is practiced on small landholdings to produce food enough for family needs. *Grow crops and eat crops* is the basic principle of subsistence farming without any marketing and commercialization.

1.3.1.4 Mixed Farming

Mixed farming emphasizes the use of two or more simultaneous agricultural activities on the same farm that are complementary to each other e.g., typical crop cultivation with livestock farming. Mixed farming promotes continuous cropping with certain associated benefits i.e., fields/farms under continuous production, increase in productivity, and profit, etc.

1.3.1.5 Advanced Farming

Contradictory to the traditional way of farming, advanced farming includes different practices i.e., usage of farm machinery, crop rotation, appropriate way of irrigation, usage of animal waste as manures, etc. to increase productivity at commercial scale.

1.3.1.6 Scientific Farming

Modern-day agriculture involves research and development (R&D) activities to practice crop cycling, organic cycling, growing of exotic and genetically modified crops, etc. for a particular piece of land. At this stage, agricultural activities are largely dependent upon new research findings that ultimately lead to high crop yield and livestock production to fulfill the needs of the large population.

1.3.1.7 Digital Farming

The spirit of Digital Farming (aka Digital Agriculture) encourages the use of Information and Communication Technologies (ICT) to monitor and manage various activities related to crop production and livestock farming at the (small-scale/large-scale) farmlands. Therefore, Digital Farming with the usage of technology aims to utilize all accessible data, information, knowledge, and skills for empowering the automation of sustainable progressions in agriculture. The advancements in digital agricultural systems have gone through many phases i.e., from modification of theoretical frameworks to actual implementation or exploitation of ICT technologies in the field of agriculture. There is no count of proposed and developed digital agricultural systems with the usage of diverse computing paradigms; however, here the few prominent chronological events regarding the use of technological advancements in Digital Agriculture have been mentioned:

- First Automated Land Evaluation System (ALES) framework was proposed in 1990 which was based on the original and most influential framework proposed by the Food and Agriculture Organization (FAO) in 1976. ALES is basically a computer program (or a land information system) that allows the land evaluators to build their own expert systems for land evaluation (The automated land evaluation system (ALES). http://www.css.cornell.edu/faculty/dgr2/_static/legacy_sw/ales/ales.html; Rossiter 1990).
- The Mediterranean Land Evaluation Information Systems (Micro-LEIS) was developed (in 1992) by integrating various software tools and technologies i.e., Geographic Information System (GIS), web, expert systems, statistics, neural networks, databases, etc. to evaluate the climate change impacts on land suitability (Rosa et al. 1992; Land evaluation decision support system (MicroLEIS-DSS). http://www.fao.org/land-water/land/land-governance/land-resources-planning-toolbox/category/details/en/c/1109820/).

- GIS was integrated with map interactions and expert systems to build an Intelligent System for Land Evaluation (ISLE) in 1999 (Tsoumakas and Vlahavas 1999; Kalogirou 2002). After ISLE, the adoption of GIS in agriculture was started to ease decision-making about land suitability for agriculture through the use of visual representations of data and spatial analysis (Mendas and Delali 2012)
- Web-based Agricultural Support System (WASS) was proposed (in 2004) for the realization of a complete agricultural system (Hu et al. 2004)
- The use of Mobile Cloud Computing (MCC) was introduced in agriculture (in 2013) which is fundamentally a combination of mobile and cloud computing technologies (Prasad et al. 2013)
- Influenced by cloud computing concepts of service provisioning i.e., (Software-as-a-Service (SaaS), Platform-as-a-Service (PaaS), Infrastructure-as-a-Service (IaaS)), the concept of Agriculture-as-a-Service was implemented in 2017 (Gill et al. 2017)

1.3.2 Agricultural Revolutions

Taking advantage of the benefits of Industrial Revolutions, the history of transformations of agricultural systems from traditional approaches to modern-day approaches can also be seen as five types of revolutions or generations (known as Agriculture 1.0 to Agriculture 5.0) (Association 2017; Zambon et al. 2019; Liu et al. 2020; Ragazou et al. 2022; Baryshnikova et al. 2022) as shown in Fig. 1.3.

1.3.2.1 Agriculture 1.0

Agriculture 1.0 represents the usage of traditional ways of framing with indigenous cultivation tools i.e., sickle, scythe, pitchfork, hatchet, spade, axe, shovel, trowel, hoe, fork, rake, etc. In this era, farming was mostly local, small-scale, and labor-intensive using animal power with low production.

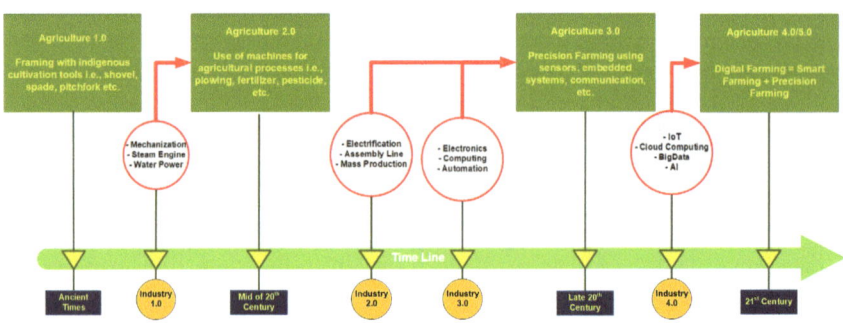

Fig. 1.3 Correspondence between agriculture revolutions and industrial revolutions

1.3.2.2 Agriculture 2.0

Agriculture 2.0 represents the era when farmers considered the advantages associated with the Industrial Revolution (Industry 1.0 from the 1780s to 1870s) and started the use of machines in agriculture for farmland preparation, sowing, plowing, irrigation, fertilization, pesticide usage, weeding, and harvesting, etc. to increase agricultural productivity with less effort (low labor-intensive). This has been considered the era of industrial agriculture or agribusiness.

1.3.2.3 Agriculture 3.0

Agriculture 3.0 era representing the usage of advancements of technological/electric and electronic/computing products of Industry 2.0 and Industry 3.0 revolutions. Technological/electric innovations of Industry 2.0 include industry electrification, assembly-line production systems, combustion engine, transportation industry, etc. that promotes the development of connected markets and shipment of agricultural products for long distance supplies. Advancement in electronics, computing, information, and communication technologies (i.e., developments of embedded systems, progressive software engineering models, global navigation system availability, data management technologies, cutting-edge wireless communication technologies) with renewable energy and power sources (hydropower, wind power, photovoltaic power) in Industry 3.0 revolution ameliorate manufacturing and automation capability. Usage of all these innovations and advancements in the agricultural sector increases the production of agricultural products (crops and livestock) through the

- replacement of small-scale farming systems with large-scale farming systems and
- establishment of precision agriculture with the availability of different applications for monitoring of outdoor farmland and indoor (greenhouse) activities and guidance farming systems.

1.3.2.4 Agriculture 4.0

Agriculture 4.0 is the era of ICT-based smart farming through the use of various boosted computing/communication technologies and paradigms of the Industry 4.0 revolution (i.e., low-cost sensors, low-cost microprocessors, high-speed wired/wireless communication and networking systems, cloud storage, BigData analytics, image processing, Artificial Intelligence (AI), drones, satellites, etc.) to assist farming with the gathering, transmission, storage, and automated analysis of agriculture-related information.

1.3.2.5   Agriculture 5.0

Agriculture 5.0 aims to apply ICT technologies introduced in Agriculture 4.0 in a way to reduce the environmental impacts of agriculture along with the solution to social and political problems of agricultural systems. Mainly it emphasizes the combining of green renewable energy sources with ICT technologies while reflecting a shift from a focus on high productivity to a focus on environmental and societal wellbeing.

Although, we are now in the era of Agriculture 5.0 (or Digital Agriculture); however, all these incarnations of agricultural revolutions (from 1.0 to 5.0) are in practice at various geographical locations of the same and/or different continents.

1.4   Digital Agriculture

The conception of digital agriculture is based on the building of automated integrated agricultural systems, which deal with data collection, transmission, and processing along with the digitalization of farm machinery control through computing, communication, and network technologies to monitor all agricultural activities (Tang et al. 2002). Hence, beyond the mere collection of data, digital farming emphasizes the use of actionable intelligence to create meaningful added value from available data (Association 2017; Diamond 2020). Digital agriculture automates all aspects or phases of the agriculture industry including planning for cropping/plantation or livestock farming, seed/breed selection, preparation of land and/or livestock sheds, planting/cultivation, and continuous monitoring of plant/tree/crop and rearing of animals along with different on-going field/farm activities such as water management, fertilizer management, insect/pest management, harvesting, post-harvest handling (i.e., transportation and storing of agricultural products, food packaging, food processing, food storage, and safety or preservation, food distribution, food quality management, and food marketing). For optimized productivity and handling of all processes at these phases of agriculture, data or information is the main crux for digital agriculture as all agriculture industry stakeholders (i.e., farmers, ranchers, extension workers, researchers, food suppliers, etc.) demand complete, accurate, in-time/on-time information for optimized decision making. Moreover, information availability and accessibility are required to be cost-effective, well-protected, and in a user-friendly form (Mahant et al. 2012). In simple words, digital agriculture provides solutions to increase the production of agricultural products by making farming practices more controlled, precise, and accurate through the use of advanced ICT technologies that help to reduce the production cost and impact of environmental factors.

1.4.1 Driving Factors for Digital Agriculture

Agriculture in the twenty-first century has been facing three main challenges i.e., impacts of extreme weather and climate variability, the rapid growth of the world's population, and the decrease of agricultural lands (which are capable for farming of crops and pastures) due to urbanization (Huang et al. 2020). According to United Nations (UN) report, the world population is projected to reach 9.7 billion in 2050, and 11.2 billion by 2100 (Cohen 2003; Gu et al. 2021) as shown in Fig. 1.4a. On the other hand, due to various environmental factors (i.e., climate change, topography, temperature, soil quality, etc.), the decreasing of agricultural land around the globe (shown in Fig. 1.4b) is further exacerbating the feeding issues of such a huge population (Zhang et al. 2018; Fao 2009; Alexandratos and Bruinsma 2012).

Along with the rapid population growth and scarcity of agricultural land, there are certain other key factors i.e., demand for remote monitoring/management, higher productivity, cost reduction, and minimization of environmental effects can also be considered as the driving factors for the usage of digital technology in agriculture. These key drivers of technology usage in agriculture have been categorized as higher yield, resource optimization, automation, and climate effects (Ayaz et al. 2019). In the modern era, the agriculture sector (including all branches and related disciplines) is highly inclined towards the adaptation of digital agriculture/farming with the usage of advanced ICT technologies, especially those that are part of the Internet of Things (IoT) technology. The stack of IoT technologies consists of sensors, communication protocols, cloud-based data storage, analytic services, actuators, etc., and aims to utilize all accessible information, knowledge, and skills for empowering the automation of sustainable progressions in agriculture (Brewster et al. 2017).

(a)   **(b)**

Fig. 1.4 **a** World population (*Source* Gu et al. 2021), **b** Percentage of used agricultural land (*Source* Food and Agriculture Organization of the United Nations 2024)

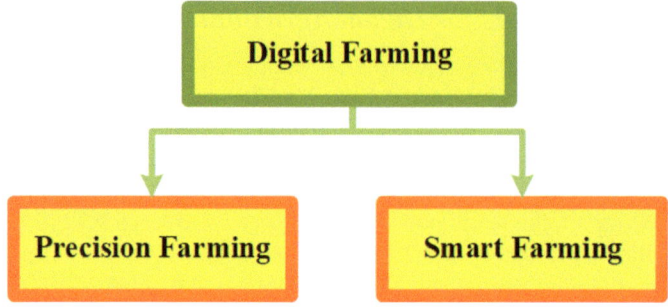

Fig. 1.5 Types of digital farming

1.4.2 Types of Digital Agriculture

In essence, digital farming is the integration of two advanced agricultural farm management concepts i.e., precision farming and smart farming as shown in Fig. 1.5. Therefore, the German Agricultural Society [Deutsche Landwirtschafts-Gesellschaft (DLG) (Beriya 2020)] defines digital farming as *"consistent application of the methods of precision farming and smart farming, internal and external networking of the farm and use of web-based data platforms together with BigData analyses"* (Griepentrog et al. 2018; German agricultural society (Deutsche Landwirtschafts-Gesellschaft (DLG). https://www.dlg.org/en/). The digital infrastructure consisting of low-cost miniature electronics devices (sensors), automated embedded hardware, advanced telematics, short-range wireless devices, broadband Internet, cloud computing models, BigData storage, drone technology, intelligent devices, efficient data processing, and analysis platforms ensures higher production along with additional data security and protection.

1.4.2.1 Precision Farming

Precision Farming (aka Precision Agriculture) inspires the usage of technology in (small/large-scale) farming to monitor and manage various activities of crop fields and livestock. The International Society of Precision Agriculture defines precision agriculture as *"a management strategy that gathers, processes and analyzes temporal, spatial and individual data and combines it with other information to support management decisions according to estimated variability for improved resource use efficiency, productivity, quality, profitability and sustainability of agricultural production"* (International society of precision agriculture (ISPA). https://www.ispag.org/about/definition). Precision agriculture in European Parliament report has been defined as *"a modern farming management concept using digital techniques to monitor and optimize agricultural production processes"* (Schrijver et al. 2016). According to this definition, the vital theme of precision agriculture revolves around optimization. For example, precision agriculture promotes the measuring of soil nutrients' variations in a field to apply fertilizer amounts adaptively to different parts of the agriculture field rather than applying an equal amount over an entire field. Precision farming also known as Site-Specific Crop Management (SSCM) (Taylor and Whelan 2005) has the basic objective of the promotion of a Decision Support System (DSS) for farm management to optimize farm production while preserving farm resources. The use of information technology in DSS decreases the management cost and increases the efficiency of the agricultural farm for the growing of crops and raising livestock. The first wave of the precision agricultural revolution reflects the usage of satellite imagery, weather prediction, variable-rate fertilizer application, etc. The second wave considers the aggregation of soil data, topographical mapping, and machine data for the precise cultivation of crops (Kukutai 2016).

1.4.2.2 Smart Farming

Unlike precision agriculture, smart farming does not specifically deal with the precise measurements and monitoring of any individual crop/field features or any individual livestock animals, respectively. On the other hand, smart farming (an advancement of precision farming) has been considered as the application of information technologies (efficient data storage and analysis techniques) to support decision-making on collected data in a smart way. Smart farming not only encompasses in-field management tasks but is based on the integration of various computing and ICT technologies (sensing, communication, cloud computing, etc.) to deal with the collection, processing, and evaluation of different kinds of agricultural data i.e., weather, climate, soil condition, crop status, terrain, manpower, resource usage, (Raja and Vyas 2019; Rehman 2015; Balafoutis et al. 2020; Walter et al. 2017; Bacco et al. 2018; Lytos et al. 2020), etc. Although, digital, precision, and smart farming/agriculture terms have been explained as distinct concepts; however, in literature sometimes these terms have been used interchangeably.

1.5 Digital Agriculture Framework

Digital agriculture is the heart of agricultural informatization and it is considered an agriculture information technology system with advanced high-tech support that is used to digitalize all the aspects of farming systems i.e., plowing, sowing, irrigation, fertilization, protection, yield prediction, harvesting, storing, monitoring, and management of agricultural phenomena and/or agricultural products. Based on the information required at different levels of agricultural administration, the digital agriculture framework covering all aspects of agricultural farming (through the use of sensing, communication, networking, information, and computing technologies) can be explained as a framework consisting of three layers i.e., Information Basis Level (IBL), Integral Application Level (IPL), and Integral Decision-Making Level (IDL) as shown in Fig. 1.6 (Liang et al. 2003). The IBL deals with the practice of technology standards under defined rules, laws, and regulations to support the infrastructure for all temporal and spatial agricultural information collection. The IPL of the digital agricultural framework includes different digital agricultural systems that can be divided into various government-oriented and public-oriented application systems as shown in Fig. 1.6. The IDL consists of administrative-level trans-regional and trans-professional application systems. These three levels of the digital agricultural framework encompass all application domains that are required for the successful realization of digital agriculture.

Questions

Q1.1: Agriculture has been considered an art as well as a science. Do you think it is the same for Digital Agriculture?

Q1.2: How Precision Farming is different from Smart Farming?

Q1.3: Describe the key driving factors for Digital Agriculture.

Q1.4: Considering the discussion on Agricultural Revolutions in this chapter explain how Agriculture 5.0 is different from Agriculture 4.0.

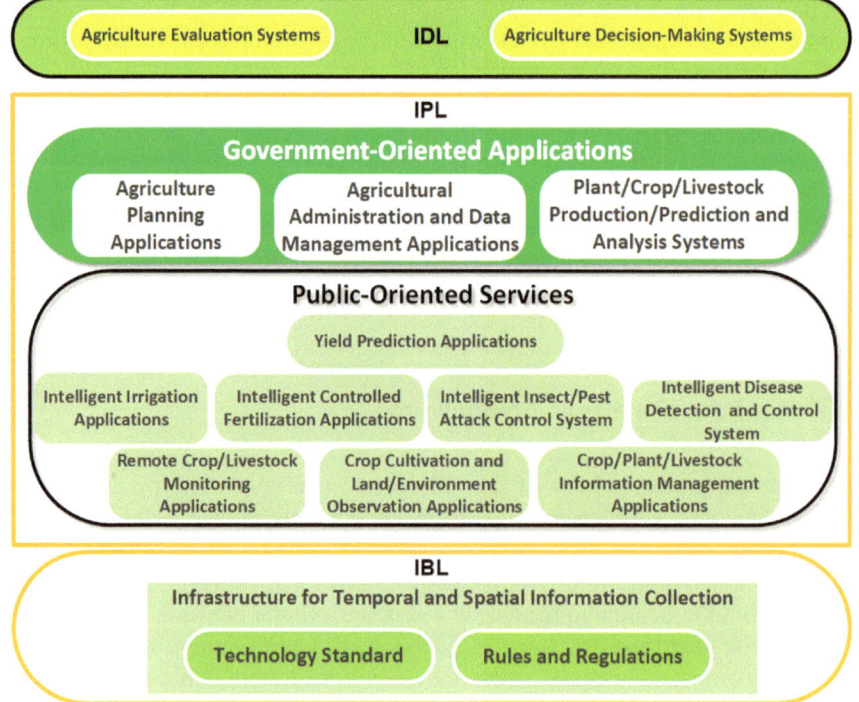

Fig. 1.6 Digital agriculture framework

Q1.5: Explain with examples how digital technologies can support environmentally friendly farming practices.

References

Alexandratos N, Bruinsma J (2012) World agriculture towards 2030/2050: the 2012 revision

Ayaz M, Ammad-Uddin M, Sharif Z, Mansour A, Aggoune EHM (2019) Internet-of-things (IoT)-based smart agriculture: toward making the fields talk. IEEE Access 7:129551–129583

Bacco M et al (2018) Smart farming: opportunities, challenges and technology enablers. In: Proceedings of the 2018 IoT vertical and topical summit on agriculture-tuscany (IOT tuscany). IEEE, pp 1–6

Balafoutis AT, Evert FKV, Fountas S (2020) Smart farming technology trends: economic and environmental effects, labor impact, and adoption readiness. Agronomy 10(5):743

Baryshnikova N, Altukhov P, Naidenova N, Shkryabina A (2022) Ensuring global food security: transforming approaches in the context of agriculture 5.0. In: IOP conference series: earth and environmental science. IOP Publishing, p 032024

Beriya A (2020) Precision agriculture to digital agriculture: a literature review

Brewster C, Roussaki I, Kalatzis N, Doolin K, Ellis K (2017) IoT in agriculture: designing a Europe-wide large-scale pilot. IEEE Commun Mag 55(9):26–33

Chandrasekaran B, Annadurai K, Somasundaram E (2010) A textbook of agronomy. New Age International Limited

Cohen JE (2003) Human population: the next half century. Science 302(5648):1172–1175

De la Rosa D, Moreno J, García LV, Almorza J (1992) MicroLEIS: a microcomputer-based Mediterranean land evaluation information system. Soil Use Manag 8(2):89–96

Diamond MSE (2020) Digital farming. http://www.foeeurope.org/sites/default/files/gmos/2020/foee-digital-farming-paper-feb-2020.pdf

E. A. M. Association (2017) Digital farming: what does it really mean. In: Position paper, CEMA

Fao U (2009) Global agriculture towards 2050. FAO, Rome

Food and Agriculture Organization of the United Nations (2024) FAO—with major processing by our world in data. In: Change in agriculture area with respect to one decade back—FAO . Food and Agriculture Organization of the United Nations, land, inputs and sustainability: land use [original data]. https://ourworldindata.org/land-use

German agricultural society (Deutsche Landwirtschafts-Gesellschaft (DLG)). https://www.dlg.org/en/

Gill SS, Chana I, Buyya R (2017) IoT based agriculture as a cloud and big data service: the beginning of digital India. J Organ End User Comput 29(4):1–23

Griepentrog WH, Uppenkamp N, Horner R (2018) Digital agricultute, a DLG position paper. DLG e.V., Frankfurt

Gu D, Andreev K, Dupre ME (2021) Major trends in population growth around the world. China CDC Weekly 3(28):604

Harris DR, Fuller DQ (2014) Agriculture: definition and overview. Encycl Global Archaeol 12:104–113

Hu Y, Quan Z, Yao Y (2004) Web-based agricultural support systems. In: Workshop on web-based support systems, pp 75–80

Huang K et al (2020) (2020) Photovoltaic agricultural internet of things towards realizing the next generation of smart farming. IEEE Access 8:76300–76312

International society of precision agriculture (ISPA). https://www.ispag.org/about/definition

Kalogirou S (2002) Expert systems and GIS: an application of land suitability evaluation. Comput Environ Urban Syst 26(2–3):89–112

Kukutai A (2016) Can digital farming deliver on its promise? PrecisionAg 28

Land evaluation decision support system (MicroLEIS-DSS). http://www.fao.org/land-water/land/land-governance/land-resources-planning-toolbox/category/details/en/c/1109820/

Liang Y, Lu XS, Zhang DG, Liang F, Ren ZB (2003) Study on the framework system of digital agriculture. Chin Geogr Sci 13(1):15–19

Liu Y, Ma X, Shu L, Hancke GP, Abu-Mahfouz AM (2020) From industry 4.0 to agriculture 4.0: current status, enabling technologies, and research challenges. IEEE Trans Ind Inform

Lytos A, Lagkas T, Sarigiannidis P, Zervakis M, Livanos G (2020) Towards smart farming: systems, frameworks and exploitation of multiple sources. Comput Netw 172:107147

Mahant M, Shukla A, Dixit S, Patel D (2012) Uses of ICT in agriculture. Int J Adv Comput Res 2(1):46

Mendas A, Delali A (2012) Integration of multicriteria decision analysis in GIS to develop land suitability for agriculture: application to durum wheat cultivation in the region of Mleta in Algeria. Comput Electr Agricult 83:117–126

National geographic society. https://education.nationalgeographic.org/resource/the-art-and-science-of-agriculture/

Oxford lexico dictionary. https://www.lexico.com/en/definition/agriculture

Prasad S, Peddoju SK, Ghosh D (2013) AgroMobile: a cloud-based framework for agriculturists on mobile platform. Int J Adv Sci Technol 59:41–52

Ragazou K, Garefalakis A, Zafeiriou E, Passas I (2022) Agriculture 5.0: a new strategic management mode for a cut cost and an energy efficient agriculture sector. Energies 15(9), 3113

Raja L, Vyas S (2019) The study of technological development in the field of smart farming. In: Smart farming technologies for sustainable agricultural development: IGI Global, pp 1–24

Rehman AU (2015) Smart agriculture: an approach towards better agriculture management. Omics Group

Rossiter DG (1990) ALES: a framework for land evaluation using a microcomputer. Soil Use Manag 6(1):7–20

Schrijver R, Poppe K, Daheim C (2016) Precision agriculture and the future of farming in Europe. Sci Technol Opt Assess

Tang S, Zhu Q, Zhou X, Liu S, Wu M (2002) A conception of digital agriculture. IEEE Int Geosci Remote Sens Symp 5:3026–3028

Taylor J, Whelan B (2005) A general introduction to precision agriculture. Australian Center for Precision Agriculture

The automated land evaluation system (ALES). http://www.css.cornell.edu/faculty/dgr2/_static/legacy_sw/ales/ales.html

Thrall PH, Bever JD, Burdon JJ (2010) Evolutionary change in agriculture: the past, present and future. Evolut Appl 3(5–6):405

Tsoumakas G, Vlahavas I (1999) ISLE: an intelligent system for land evaluation. Proceed ACAI 99:26–32

Walter A, Finger R, Huber R, Buchmann N (2017) Opinion: smart farming is key to developing sustainable agriculture. Proceed Natl Acad Sci 114(24):6148–6150

Zambon I, Cecchini M, Egidi G, Saporito MG, Colantoni A (2019) Revolution 4.0: industry versus agriculture in a future development for SMEs. Processes 7(1), 36

Zhang L, Dabipi IK, Brown WL (2018) Internet of things applications for agriculture. In: Hassan Q (ed) Internet of things A to Z: technologies and applications. Wiley, Amsterdam

Chapter 2
Digital Agriculture Ecosystem

2.1 Learning Objectives

After studying this chapter, students will be able to

- describe the concepts of ecosystem, agricultural ecosystem, digital ecosystem, and digital agricultural ecosystem
- identify the major components of the digital agricultural ecosystem
- explain the fundamental details relevant to the digitalization (or processing) of agricultural data at different phases of agricultural products in terms of collection, transmission, storage, and analysis
- illustrate the placement of enabling digital technologies (required to automate the monitoring and management of agricultural operations) at different layers of IoT-based digital agriculture ecosystem

2.2 Ecosystem

An ecosystem is a system in a specific environment where living (plants, animals, microbes) and non-living entities (weather, climate, soil, water, landscape, etc.) interact with each other to create a bubble of life (National geographic: ecosystem definition. https://education.nationalgeographic.org/resource/ecosystem/). In an ecosystem, all living entities rely on each other to support life. The feeding relationships between organisms of an ecosystem can be described as a food chain. In the food chain, plants are dependent on various environmental factors (i.e., sunlight, water, soil nutrients, air, etc.) and produce food for animals and humans. The wastes and organic matter produced by plants, and humans are broken down by microbes to soil humus which are then available for the plants to use. The basic steps in a food chain are shown in Fig. 2.1.

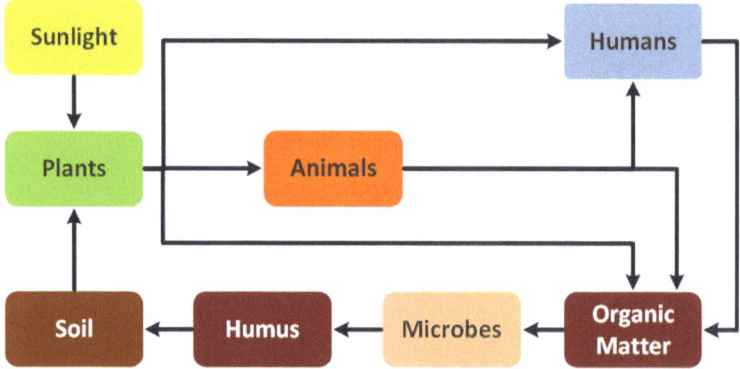

Fig. 2.1 Classical food chain steps in an ecosystem

Ecosystems can be broadly classified into two types of ecosystems i.e., Natural Ecosystems and Artificial (or Managed) Ecosystems. Natural ecosystems are complex in functioning and usually consist of many species of plants and animals. Forests, oceans, mountains, grasslands, etc. are a few examples of natural ecosystems. On the other hand, artificial ecosystems are modified and controlled by humans. One of the examples of artificial ecosystems is the Agricultural Ecosystem.

2.3   Agricultural Ecosystem

An agricultural ecosystem is an artificial ecosystem that is created, managed, and controlled by humans (Guidotti 2015). To put it differently, agricultural ecosystems are simplified natural ecosystems because they have relatively few species of plants and animals that the farmer wants to grow and rear, respectively. The structure of the agricultural ecosystem is characterized by the interconnected network of living organisms and their environment that work together to improve agricultural productivity and sustainability. Modern agricultural ecosystems are mainly created, managed, and controlled by the use of advanced technologies that are part of Digital Ecosystems.

2.4   Digital Ecosystem

In recent times, digital technologies have become ubiquitous with a promise to transform our daily lives, societies, and economy. Modern digital tools, devices, and systems in an organized interconnected form of grouping build a digital ecosystem. By definition, a digital ecosystem is an adaptive, scalable, and sustainable system with a group of interconnected digital resources forming a single information environment to promote members' interaction when no hard functional ties exist between them (Barykin et al. 2020). In simple words, a digital ecosystem comprises a network of systems and respective stakeholders that use digital technologies to interact with one another to pursue mutual economic opportunities. Like other fields, the agriculture sector is highly inclined towards the adoption of digital technologies especially those that are part of the IoT technology stack. The IoT technology stack mainly comprises sensors, communication devices, data storage resources, analytic services, software applications, etc., and provides the basis of the Digital Agricultural Ecosystem.

2.5 Digital Agricultural Ecosystem

A Digital Agricultural Ecosystem is an adaptive, scalable, and sustainable system established with a group of interconnected digital resources that offer an information environment to assist the automation of agricultural operations while promoting interaction among agricultural stakeholders when no firm collaboration and functional bonds exist between them (Kuldeep Singh 2024). The key objective of the digital agricultural ecosystem is to produce a higher quantity of high-quality agricultural yield through the adoption of digital technologies in a sustainable way. Therefore, along with the benefits of reduced labor costs and increased income, digital agriculture ecosystems empower farmers and farming communities to make effective decisions to increase and improve the quantity and quality of their agricultural outputs, respectively. The four major components of the digital agriculture ecosystem are agricultural stakeholders, agricultural machinery, agricultural operations, and digital technologies.

Agricultural Stakeholders include farmers, ranchers, seed providers, pesticide and fertilizer companies, agricultural consultants, retailers, advisors, agricultural extension workers, aggregators, distributors, marketers, transporters, animal feed companies, veterinarians, agricultural credit and financial institutions, etc.

Agricultural Machinery includes combine harvesters, forage harvesters, tractors, and tractor attachments (plows, harrows, seeders, fertilizer spreaders, sprayers, mowers, cultivators, balers, wagons or trailers, orchard cabs, etc.)

Agricultural Operations at different phases of crop and animal production have been discussed below (and shown in Fig. 2.2) (Aker et al. 2016) i.e.,

- *Pre-cultivation*—land preparation, planting, sowing, etc.
- *Cultivation*—irrigation, fertilization, spraying pesticide, weeding, plant counting, yield prediction, monitoring and management of crops, crop fields, and farm resources, etc.
- *Harvesting*—crop harvesting along with labour and machinery management.
- *Post-harvest handling*—transportation and storage of agricultural products, food packaging, food processing, food safety or preservation, food distribution, food quality management, and food marketing, etc.

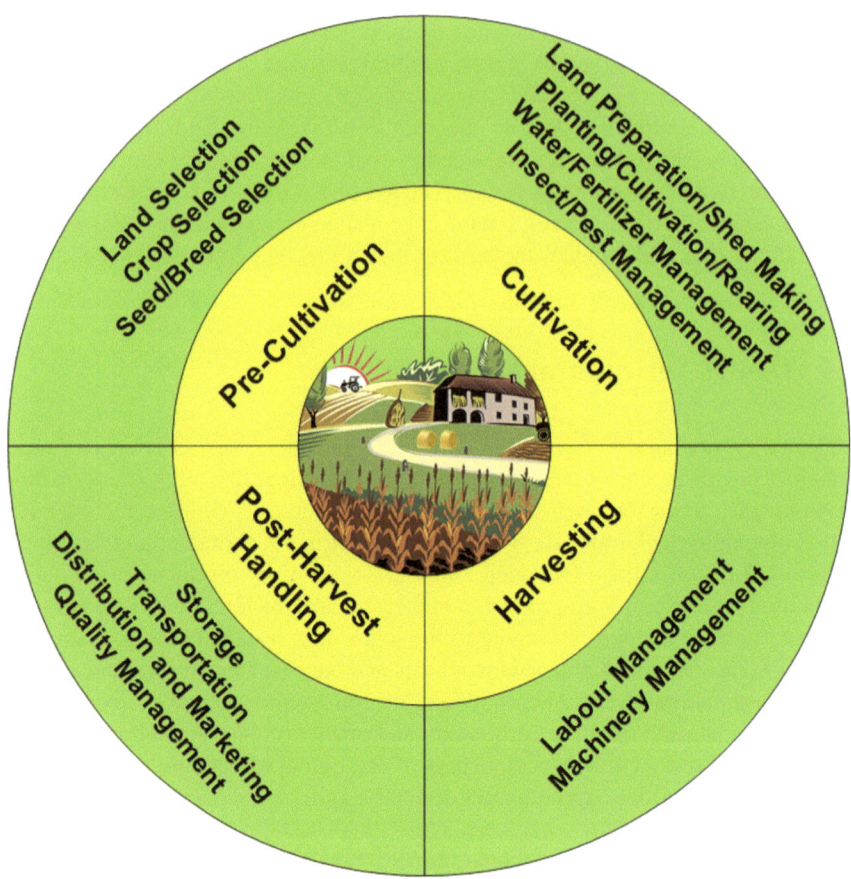

Fig. 2.2 Agricultural operations at different phases of agricultural product

Digital Technologies include all electronic devices and computing techniques that are mainly the part of IoT technology stack including sensors, drones, wired and wireless communication, data/cloud storage, and data analysis using AI techniques (including machine learning and deep learning). The use of IoT technologies assists in continuous real-time monitoring and automated management of agricultural operations and farm machinery at the agricultural farmlands. It also reduces farmer's workload and improves the efficient utilization of farm resources through cost-effective data collection, storage, processing, and analysis to predict and increase farm's expected agricultural productivity (Udutalapally et al. 2020). As data is the core of any digital ecosystem and therefore it becomes obvious to understand different phases of data processing to fully grasp the working details of other functional components of the digital ecosystem.

2.6 Data Processing in Digital Agriculture Ecosystem

Digital agricultural ecosystems work on the principles of IoT that are fundamentally based on the network of digital resources, smart devices, digital technologies, and computing techniques that increase the efficiency of agricultural operations and ultimately improve the production and quality of agricultural products. The architecture of the IoT-based digital agriculture ecosystem is shown in Fig. 2.3. The digitalization of agricultural operations in an IoT-based digital agriculture ecosystem is accomplished in different phases of data processing i.e., Data Acquisition, Data Transmission, Data Storage, and Data Analytics (as shown in Fig. 2.3). The details associated with these data processing phases in an agricultural context have been discussed below.

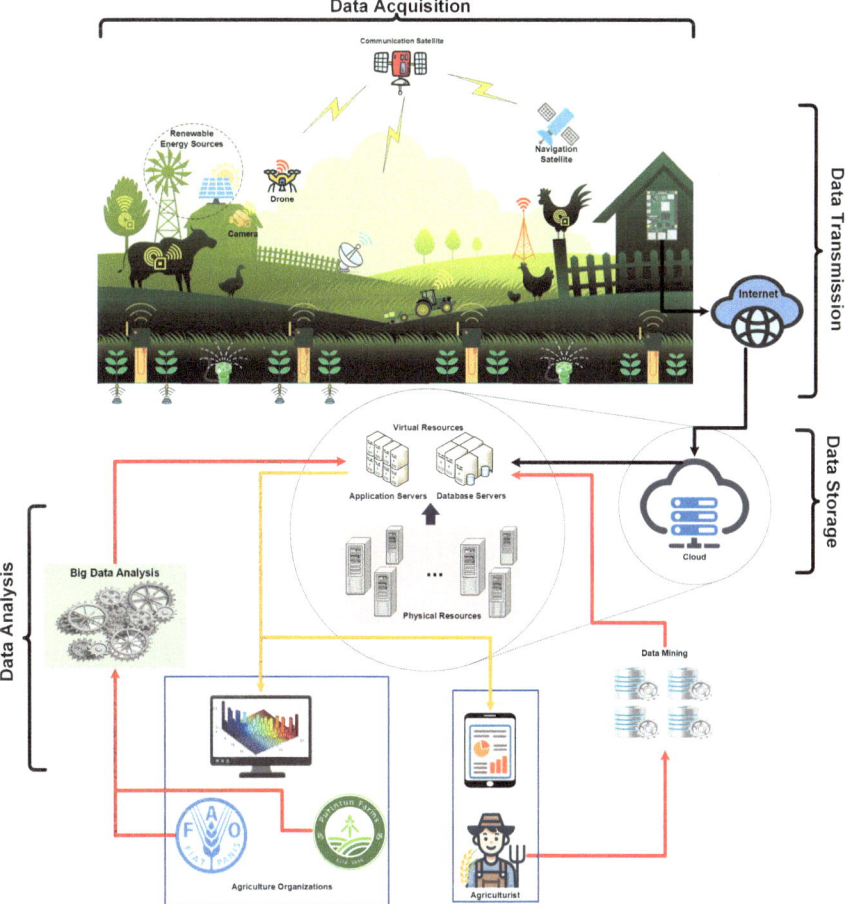

Fig. 2.3 IoT-based digital agricultural ecosystem

2.6.1 Data Acquisition

The data acquisition phase deals with the collection of various types of agricultural data i.e.,

- Data required for better crop/plant growth and management that include
    - soil moisture, temperature, texture
    - climate conditions
    - temperature and humidity
    - pest infestation and crop diseases
    - fertilization and irrigation need for crops
    - plant growth patterns

- Data required for better livestock production that include
    - animal movement
    - animal grazing patterns
    - animal rumination
    - animal health and well-being (i.e., hearth-rate, heat, stress level monitoring, etc.)

through the use of various digital technologies i.e., RFID (Radio Frequency IDentification) tags, RGB and multispectral cameras, Unmanned Aerial Vehicles (UAVs), satellites, GPS (Global Positioning System), mobile phones, and predominantly by various types of available sensors (i.e., pH sensor, gas sensor, ultraviolet sensor, rainfall sensors, motion detection sensor, soil moisture sensor, temperature sensor, humidity sensors, barometric pressure sensors, etc.). Relevant details regarding the basic functionality of these devices and their use in different agriculture scenarios have been discussed in Chap. 3.

2.6.2 Data Transmission

The transmission of agricultural data (vital field parameters i.e., soil, crops, livestock, weather, etc.) from field sensors/devices (mentioned in Sect. 2.6.1) to edge routers is mainly based on wireless communication infrastructure. Based on the connectivity spectrum, various advanced wireless technologies that are part of digital agriculture ecosystems can be categorized as

- Short-Range Communication Technologies (i.e., RFID, NFC, Bluetooth, ZigBee, etc.)
- Long-Range Communication Technologies (i.e., LoRaWAN, NB-IoT, SigFox, etc.)
- Improved Local Area Communication Technologies (WiFi 6, WiFi HaLow, etc.)
- Cellular Communication Technologies (2G/3G, LTE)

- High-Band Communication Technologies (4G, 5G)
- Satellite Telecommunications

Other than efficient wireless communication infrastructure, data transfer from edge routers/gateways to large-scale data centers is generally dependent on Broadband technology. Related details regarding these technologies' essential functionality and usage in different agriculture scenarios have been elaborated in Chap. 4.

2.6.3   Data Storage

Data storage consisting of large-scale cloud data centers deals with the storage of massive amounts of data (BigData) generated by IoT-based agricultural applications. Agricultural applications offer services that cover all aspects of agriculture from planting to harvesting of crops and rearing of animals in the small-/large-scale farmlands through the use of various sensors and IoT devices. These sensors and IoT devices generate a huge amount of heterogeneous (structured, semi-structured, and unstructured) data that is usually not possible to store on locally available hardware resources. Furthermore, commonly used data management tools are unable to manage and process this immense amount of data. Therefore, Cloud Computing with the provisioning of huge IT infrastructure enables IoT systems to meet the elastic demands of agriculturists and agricultural organizations in terms of storage and computing resources (Al-Fuqaha et al. 2015). Details regarding the implementation of cloud computing and related technologies (i.e., Fog/Edge Computing) in agricultural scenarios have been explained in Chap. 5.

2.6.4   Data Analytics

Data Analytics deals with the analysis of stored data through the usage of different computing technologies i.e., image/video processing, computer vision, pattern recognition, artificial (machine/deep learning) approaches, audio conversion, natural language processing, etc. The accuracy of data analysis in agriculture farming ensures high productivity through improved operational efficiency. Different types of data analytical techniques improve the quantity and quality of agricultural products (Elijah et al. 2018; Wolfert et al. 2017) by

- optimizing the use of resources by identifying environmental factors (e.g., humidity, temperature) that ultimately reduce the operational and management cost of an agricultural farm
- predicting extreme conditions i.e., flood, drought, disease, etc., to take preventive measures instantly
- detecting pests and crop diseases at early stages of crop development to take in-time protective measures

- assisting (private and government-level) agricultural departments, organizations, and companies in better decision-making for short/medium/long term planning, policies, and marketing strategies.

Implications of data analysis techniques have been discussed in detail in Chap. 6.

2.7 Layered Architecture of IoT-Based Agricultural Ecosystem

The enabling environment of an IoT-based agricultural ecosystem can be represented as layered architecture as shown in Fig. 2.4. The layered architecture consisting of five layers (i.e., perception, network, cloud, analysis, and application) establishes the foundation that helps to better understand the placement of digital technologies along with their functional correspondence to different stages of data processing related to various agricultural operations and activities.

The *Perception Layer* utilizing various devices i.e., RFID (Radio Frequency IDentification) tags, sensors, RGB and multispectral cameras, Unmanned Aerial Vehicles (UAVs), satellites, GPS (Global Positioning System), and mobile phones is responsible for data collection of various environmental factors and agricultural activities at all phases (from pre-cultivation to post-harvesting) of agricultural products. Mostly, the sensed data is transmitted to microcontrollers or microprocessors that in turn are responsible for transmitting collected data (with corresponding actions) to the cloud for storage.

The *Data Communication (or Network) Layer* consisting of various communication technologies i.e., ZigBee, Bluetooth, WiFi, LoRa, 4G/5G, Broadband, etc. is responsible for transmitting (sensed or captured) data from the perception layer to the storage layer. These wired/wireless modules are mostly embedded in microcontrollers and microprocessors to facilitate the forwarding of collected data to the cloud resources.

The *Data Storage (or Cloud) Layer* is responsible for the storage of huge volumes of (structured, semi-structured, and unstructured) agricultural data collected from different sources at different stages of agricultural operations. Structured data has a standardized format typically tabular format with rows and columns (defining clear data attributes) and is effectively processed by computing techniques for insights. Examples of structured data include data stored in relational databases, excel files, web forms, etc. Semi-structured data is not fully structured data as it lacks completeness of information and does not exist in tabular form. However, semi-structured datasets include metadata (with tags and markers) that is helpful for analysis. Examples of semi-structured data are emails, zipped files, JSON (JavaScript Object Notation), CSV (Comma-Separated Values) files, etc. The unstructured datasets have an internal structure but lack a predefined schema or format. Examples of unstructured data include audio/video files, photo files, text files, etc.

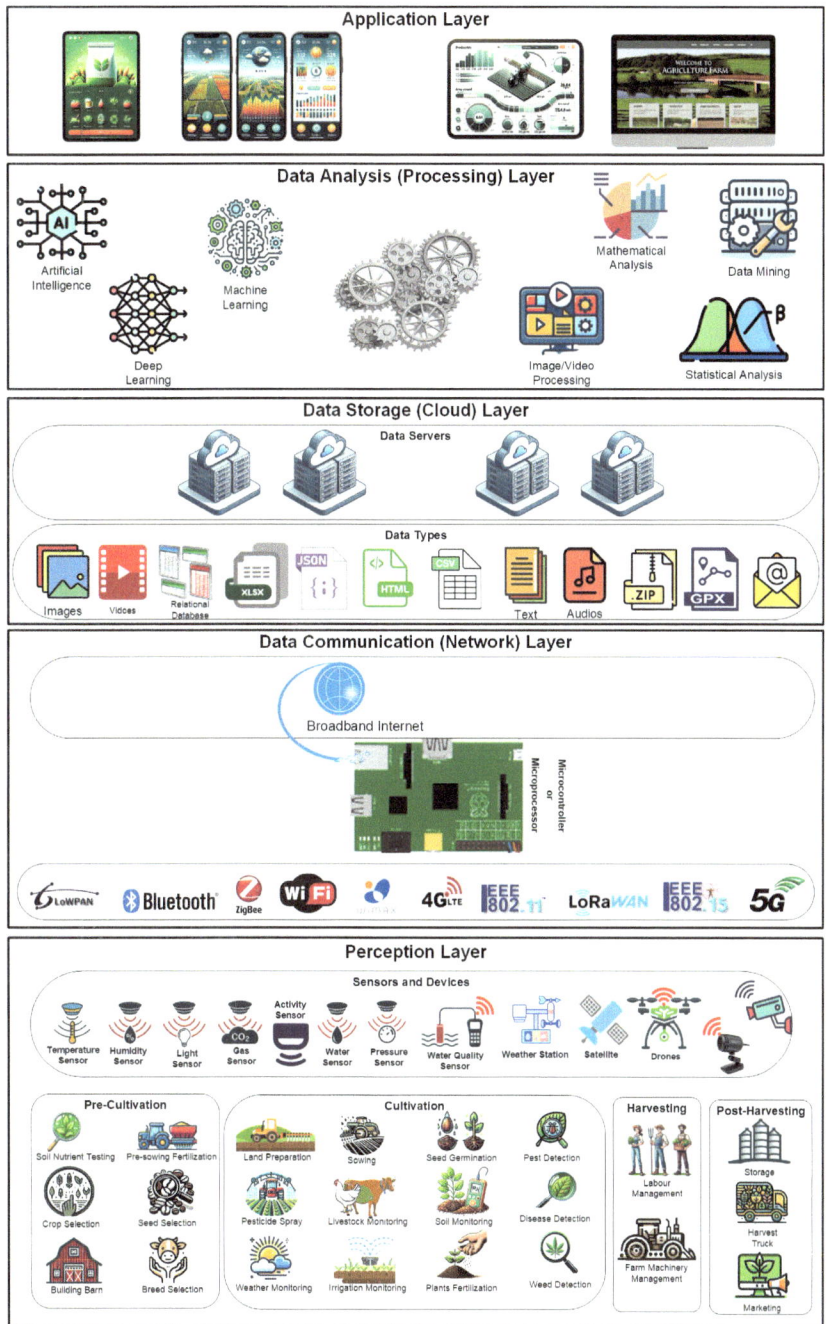

Fig. 2.4 Layered architecture of IoT-based Agricultural Ecosystem

The **Data Analysis (Processing) Layer** is responsible for assessing data stored in the cloud through intelligent computational techniques that provide useful insight and help businesses in decision-making. This layer mainly consists of two components i.e., data preprocessing and data analysis. Data preprocessing involves data cleaning to find errors and inconsistencies in data. The data analysis component, on the other hand, involves the identification of various valuable patterns and relationships by applying computational intelligence using various statistical, mathematical, machine/ deep learning, and natural language processing techniques.

The **Application Layer** provides services with user-friendly interfaces to interact and visualize the results of data analysis. Typically, the interaction through user-friendly interfaces allows users to monitor, configure, and control various indoor/ outdoor environmental factors that are part of the IoT-based agricultural system. The visualization techniques at this layer enable agriculturists to drill down and focus on more detailed views of analysis results on available data through interactive charts.

Questions

Q2.1: Write the name of agricultural stakeholders and how do these agricultural stakeholders get benefits from the digitalization of agricultural operations?

Q2.2: Describe the use of a video camera and microphone (shown in Fig. 2.4 at the perception layer of an IoT-based Digital Agricultural Ecosystem) with the help of an example.

Q2.3: Which digital technologies are involved in IoT-based agricultural systems to detect plant diseases and pest attacks in an agricultural field?

Q2.4: Non-stop monitoring of the barn/pasture environment and cows' activities on a dairy farm is required to improve the quantity and quality of milk production. To address this challenge, an IoT system based on various IoT devices and technologies is required to be implemented to assist dairy farmers in their everyday decisions. Considering the implementation perspective of this type of IoT-based dairy farming system.

a. Draw a clear architecture diagram for this system
b. Label all digital components shown in this architecture diagram
c. Write the names of different types of sensors, IoT devices, and advanced technologies that can be used for the successful actualization of this system

Q2.5: Write the types of sensors, digital technologies, and devices required to be implemented in the Tomato and Carrot Field (shown in Fig. 2.5) to automatically monitor the temperature, humidity, soil moisture, soil nutrients, rainfall, wind speed, and wind direction.

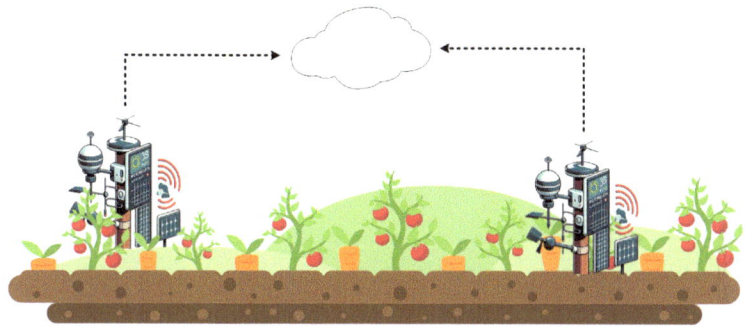

Fig. 2.5 Smart tomato and carrot field (figure for Question 2.11)

References

Aker JC, Ghosh I, Burrell J (2016) The promise (and pitfalls) of ICT for agriculture initiatives. Agricult Econ 47(S1):35–48

Al-Fuqaha A et al (2015) Internet of things: a survey on enabling technologies, protocols, and applications. IEEE Commun Surv Tutor 17(4):2347–2376

Barykin SY et al (2020) Economics of digital ecosystems. J Open Innov Technol Market Compl 6(4):124

Elijah O et al (2018) An overview of Internet of Things (IoT) and data analytics in agriculture: benefits and challenges. IEEE Internet Things J 5(5):3758–3773

Guidotti TL (2015) Artificial ecosystems. In: Health and sustainability: an introduction. Oxford University Press, Oxford

Kuldeep Singh PK (2024) Software testing techniques. Wiley Press, Amsterdam

National geographic: ecosystem definition. https://education.nationalgeographic.org/resource/ecosystem/

Udutalapally V et al (2020) sCrop: a internet-of-agro-things (IoAT) enabled solar powered smart device for automatic plant disease prediction. arXiv preprint arXiv:2005.06342

Wolfert S et al (2017) Big data in smart farming: a review. Agricult Syst 153:69–80

Chapter 3
Data Acquisition in Digital Agriculture

3.1 Learning Objectives

After studying this chapter, students will be able to

- describe the main enabling technologies of agricultural data acquisition
- illustrate the usage of various data acquisition technologies (i.e., RFID, Sensors, RFID-based Sensors, smartphones, and remote sensing platforms) in agricultural fields.
- explain the role of Development Boards in the agricultural data acquisition phase of IoT-based Digital Agriculture Systems.

3.2 Agricultural Data Acquisition

The digitalization of conventional field management practices has converted agriculture to data-driven agriculture, which leads to the more precise, timely, and cost-effective management of crops and livestock in diverse agricultural environments (including both indoor and outdoor). However, in data-driven agriculture, acquiring field or farm data is one of the most challenging tasks and directly affects the overall efficiency of digital agriculture systems. Reliable data collection of soil, crops, and field's environment parameters i.e., temperature, humidity, light, water, air, greenhouse gases, etc. helps to improve the overall yield. These days the use of various digital technologies i.e., Electronic Identification Systems, RFID, Sensors, Smart Phones, and Static/Remote Sensing platforms on the farms has enabled agriculturists to collect accurate and in-time field data. Two main characteristics of modern digital technologies i.e., object identification and detection of physical sensation help the collection of different types of agricultural data (Miorandi et al. 2012; Raj and Raman 2017). Considering the significance of these two characteristics, the content of this

M. A. Iqbal, *Digital Agriculture*,
SpringerBriefs in Agriculture, https://doi.org/10.1007/978-3-031-67679-6_3

Chapter includes the details regarding the components and functioning of various digital data acquisition systems that are being used in the agricultural context.

3.3 RFID Systems

Electronic identification systems encompassing various digital technologies are designed to identify objects through electronic devices. Considering the aim of efficient management and protection of plants/crops/trees and livestock, the electronic identification methods support the realization of digital agricultural systems through

- identification of plant, crop, tree, or an area in farmland/greenhouse
- identification, monitoring, and management of livestock
- information aggregation about the origin, characteristics, and phases of supply-chain management of food products, etc.

One of the most common types of modern electronic identification systems is Radio Frequency IDentification (RFID) system. RFID uses electromagnetic fields to automatically identify and track objects through tags that are attached to physical objects (Hunt et al. 2007). A typical RFID system consists of the following three components (Hunt et al. 2007) (shown in Fig. 3.1):

- An RFID Tag (also known as Transponder or Smart Label) composed of an antenna, (optional) battery, and semi-conductor chip

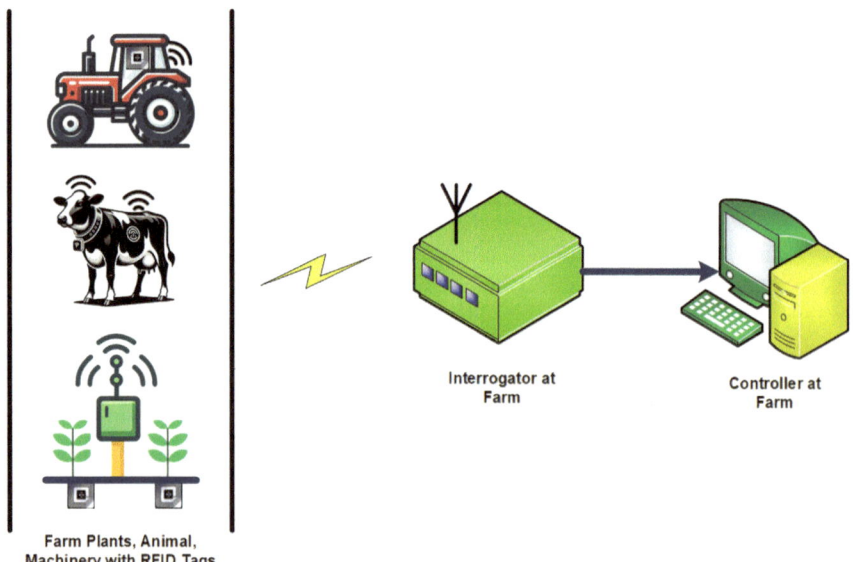

Fig. 3.1 Building blocks of RFID system at typical digital agriculture farm

- An Interrogator (also known as a Reader or read/write device) having an RF module, control module, and antenna
- A Controller (also known as a Host or Workstation with special software) to process and store required information in a database

The RFID tag and interrogator in the RFID system communicate with each other through radio waves and do not require line-of-sight. Within the transmission range, the Interrogator reads the required information (i.e., serial number, etc.) stored on the RFID tag and directs this information toward the Controller that ultimately uses this information for various purposes.

RFID tags can be of two types i.e., Active Tags (Tags having an on-board power source) and Passive Tags (Tags without an on-board power source). Active RFID Tags have greater capabilities (i.e., more memory, high transmission range, etc.) than Passive Tags but are more complex and expensive. Active Tags use battery power and can transmit information over a longer range. On the other hand, to transmit information, Passive RFID Tags can derive power from the signal received by the Interrogator. RFID usage covers a wide spectrum of application areas i.e., object identification, asset tracking, manufacturing, supply chain management, payment systems, and location identification, etc. A few examples of the implications of RFID in the agriculture domain have been described in Table 3.1 (Luvisi 2016).

Table 3.1 Common usage of RFID in agriculture

Application	Purpose
Plant management	To identify the plant, plant age, plant health, etc.
Horticultural plant markings	To get information about planting instructions, plant care, list of companion plants, etc.
Crop management	To gain valuable insights into crop health and growth patterns, etc.
Monitoring and management of farming operations	To track the origin and trace the handling of agricultural products, to track and manage farm equipment and supplies, to monitor and record the date and location of planting, fertilizer/pesticide application, automatic capturing of harvesting date, location, tree identification, etc.
Pesticide/ fertilizer management	Track the usage of pesticides and fertilizers, management of agrochemicals, impact assessment of agrochemicals, etc.
Supply chain traceability	To allow distributors to track the movement of agricultural products from farm to table ensuring food safety while monitoring the appropriate conditions for storage and transportation of these products throughout the supply chain. To track and keep a record of the origin (region), variety, year, brand, of agricultural products, etc.

3.4 Sensing Technology and Sensors

Sensing involves the originating of data from real-life objects through the use of sensors. The sensor is one of the essential components of smart things that induce the sensing capability in smart things to perceive a change in the ambient conditions occurring around their environment. Formally, a sensor can be defined in various ways such as

- In general, a device that can receive and respond to a stimulus (for example, variation in any natural phenomenon i.e., temperature, pressure, humidity, motion, position, displacement, sound, force, flow, light, chemical presence, etc.).
- In an operational sense, a sensor is a device that translates the received stimulus (a quantity, property, or condition/state of a physical object) into an electrical signal.
- In technical terms, a sensor is an electronic device responsible for producing electrical, optical, or digital data deduced from the physical environment that is further electronically transformed into useful information for intelligent devices or people.

Considering all these definitions of sensors, it can be concluded that the input of a sensor is some kind of observation (non-electric) related to the change in the physical property of an object and the output is an electrical signal in terms of variation in charge, voltage, and current. The sensor output is ultimately required to be compatible with electronic circuits. For example, through sensors, temperature and pressure can be measured by converting heat and atmospheric pressure into electrical signals, respectively. A few common sensors with generic purpose and their usage in agricultural scenarios have been discussed in Table 3.2. Moreover, few industry-standard agriculture sensors with measuring parameters are given in Table 3.3 (Fraden 2004; Shafi et al. 2019; Mathur 2020; Ayaz et al. 2019; Farooq et al. 2019; Elijah et al. 2018).

3.5 Sensor Networks

A sensor network comprises many tiny (wirelessly connected) sensors, which are distributed in an ad hoc manner in a particular environment. These tiny sensor nodes work in collaboration for the measurement and transmission of a certain type of physical phenomenon (through sink nodes) to remote databases, data warehouses, or the cloud for analysis which ultimately helps in decision-making. For the broadcast of sensed data, progressive sensor devices are equipped with different types of wireless communication and network technologies. Here, wireless communication modules (combined with the sensors) allow the transmission of heterogeneous data, and wireless networking schemes help the interconnectivity of devices for the efficient transmission of sensed data. In real-life situations, many networks are included

Table 3.2 Common sensors with generic purpose and agricultural use cases

Sensor	General purpose	Example usage in agriculture
Electromagnetic sensors	Transform an environmental quantity to be measured as induced current or voltage outputs by some form of signal or circuit parameter	Used to measure organic matter and contamination in the soil while exploiting the capability of soil particles' conductivity or accumulation of electrical charge
Humidity sensor	Sense and measure the humidity level (temperature and moisture ratio) in the air	Helps farmers to adjust the humidity level in outdoor/indoor farmlands that affects plant respiration, photosynthesis, leaf growth, pollination, etc.
Temperature sensor	Used to measure the amount of heat energy (or coldness) generated by an object or system	To measure soil, plant, and ambient temperature in outdoor farmlands or indoor farmhouses
Soil moisture sensor	Uses capacitance to measure the dielectric permittivity of the surrounding medium	Measure the water content in the soil
Load/weight sensors	Measures the load, weight, compression, tension, the pressure of/on an object	Used in the weighing system of the farm truck
Acceleration sensors	Measures acceleration caused by movement, vibration, collision, etc.	Used in farm motors and moving components to notice slight movement variations or vibration inconsistencies and helps in the prediction of required maintenance or replacement of a component or farm machinery equipment
Flow sensors	Measurements and regulation of the flow rate of liquids and gasses within pipes. Also able to detect blockages, leakage, liquid concentration, etc. in the supply pipeline	Management of flow control of water sources in the farm irrigation system
Air flow sensors	Measure the volume or mass of moving air in a channel	Used to measure soil air permeability to identify soil type, soil structure, compaction, moisture level, etc.

(continued)

Table 3.2 (continued)

Sensor	General purpose	Example usage in agriculture
Acoustic sensors	Measures the levels or intensity of sounds	Used for the • detection and monitoring of pests by their sound • detection of under-surface soil material by measuring noise upon farming tool interaction with the soil • classification of seed varieties
Field-programmable gate array (FPGA)-based sensors	FPGA as reprogrammable hardware technology helps obtain a reconfigurable sensor system	Used to measure humidity, plant transpiration, irrigation, etc.
Optical sensors	Uses the light or light reflectance phenomena by converting light rays into an electronic signal	Used to measure soil minerals, clay contents, soil moisture, organic substances, etc.
Ultrasonic ranging sensors	Use for distance measuring of a target object by emitting ultrasonic waves	Used for spray distance measurements, monitoring of overwatering/underwatering, agricultural object detection, weed detection (when combined with camera), etc.
Optoelectronic sensors	Used to produce an electrical signal proportional to the amount of light incident on its active area	Used to • perform weed mappings • differentiate plant types • detect weeds, etc.
Electrochemical sensors	Used to obtain information about the composition of a system	Used to measure soil nutrients, soil pH, salinity, etc.
Eddy covariance-based sensors	Measure three-dimensional wind velocity and vertical flux densities of CO_2, water vapor, and other greenhouse gases	Used to measure field and greenhouse gas fluxes
Soft water level-based (SWLB) sensors	Used to measure water level dynamics	Used to study the hydrological behavior of agricultural catchments by measuring natural water sources i.e., rainfall, stream flows, etc.
Light detection and ranging (LIDAR)	Measure distances by using laser light to illuminate the target area and calculate the reflection with a sensor	Used for monitoring 3D farm modeling, detecting soil type, measuring leaf area, yield forecasting by creating detailed maps of crop fields, etc.

(continued)

Table 3.2 (continued)

Sensor	General purpose	Example usage in agriculture
Telematics sensors	Supports the communication of vehicular objects	Mounted on the farm machinery and vehicles (such as tractors, harvesters, etc.) that are used for the collection of farm operations data
pH sensors	Measure acidity or alkalinity	Used to measure and monitor the amount of soil nutrients that are required for healthy plant growth
Gas sensors	Considering the consumption of infrared radiation, gas sensors are used for the identification of different types of gases	Measures the quantity of toxic gases in the greenhouse and/or in livestock farms
Infrared sensors	Used to detect and measure infrared radiation in its surrounding environment	Measure dry matter, soil nitrogen, yield, monitoring of pests, etc.
Motion detector sensor (passive infrared sensor)	Used to detect nearby motion of an object	Used in burglar alarm systems in farm fields to detect the motion of unwanted objects (i.e., thieves or animals)
Barometric pressure sensor	Used to measure changes in atmospheric pressure	Used to control water flow in the crop field based on expected rainfall by measuring the atmospheric changes

in the transmission of sensed data from its origin to the actual target device or platform. Contrary to RFID devices (which are more focused on the identification and tracking of objects with unique identifiers), sensors are able to collect more detailed and varied environmental data (Ruiz-Garcia et al. 2009). Primarily agricultural applications that can be implemented using sensor networks include the monitoring of different ongoing activities on agriculture farms (Ojha et al. 2015) i.e., monitoring the

- variation in temperature, humidity, and wind speed in the field
- probability of insect/pests occurrences
- variation in soil nutrients
- cattle's movements in grazing fields
- emission of greenhouse gases
- asset tracking
- position of farming vehicles in the field
- quality of groundwater

Two types of sensor networks i.e., Terrestrial Sensor Networks (TSNs) and Underground Sensor Networks (USNs) are particularly used in agricultural applications.

Table 3.3 Few industry standard agriculture sensors

Sensor	Parameters to measure
Gro water sensor	Soil moisture
ECH2O soil sensor	Soil moisture, soil temperature, soil conductivity
Hydra probe II soil sensor	Soil moisture, soil temperature, soil conductivity, salinity level
EC sensor (EC250)	Soil moisture, soil temperature, soil conductivity, salinity level
MP406 soil moisture sensor	Water content in soil
Pogo portable sensor	Soil temperature, soil moisture
EasyBloom 1000 plant sensor	Light, temperature, soil moisture
Xiaomi 4 in 1 plant flower care smart monitor	Light, temperature, soil moisture, soil pH
SenseH2TM hydrogen sensor	Hydrogen, plant temperature, plant wetness
LW100, leaf wetness sensor	Plant temperature, plant moisture, plant wetness
237 leaf wetness sensor	Plant temperature, plant moisture, plant wetness
Field scout CM1000TM	Chlorophyll concentration (indirectly Photosynthesis)
YSI 6025 chlorophyll sensor	Photosynthesis
TPS-2 portable photosynthesis	Photosynthesis, plant moisture, CO_2
Cl-340 hand-held photosynthesis	Photosynthesis, plant moisture, plant hydrogen level, plant wetness, CO_2, plant temperature
PTM-48A photosynthesis monitor	Photosynthesis, plant moisture, plant wetness, CO_2, plant temperature
CM-100 compact weather sensor	Air temperature, air humidity, air pressure, wind speed
HMP45C	Air temperature, air humidity
SHT71, SHT75	Air humidity, air temperature

In TCNs, small sensor nodes [mostly based on Micro-Electro-Mechanical System (MEMS) technology] are deployed above the ground surface to collect accurate data from the surrounding environment. Based on sensed data in the agricultural farm, these sensor nodes formed an ad hoc network to accomplish application requirements. For example, the automation of the irrigation system in precision agriculture demands the deployment of sensors throughout the field which eventually determine the moisture content in the soil and collaboratively make the decision about the scheduling of field irrigation. On the other hand, in USNs, the wireless sensor nodes are planted inside the soil. These sensors in the subsurface region of soil are responsible for measuring soil toxicity of surrounding environment (Rahaman and Azharuddin 2022). A typical deployment of a sensor network (with on-ground and underground sensor nodes) for agricultural applications has been shown in Fig. 3.2.

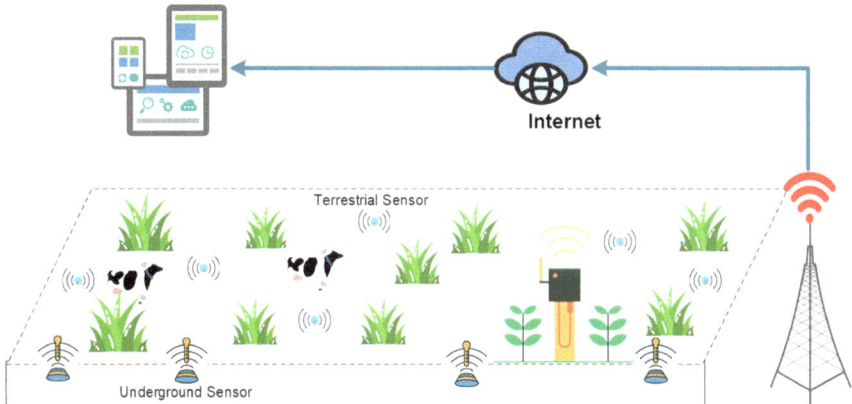

Fig. 3.2 Typical sensor network implementation (on-ground/underground) for agricultural applications

3.6 RFID-Based Sensor Networks

Currently, the integration of RFID technology with sensor nodes has been considered to support ubiquitous and pervasive computing networks. In RFID-based sensor platforms, the RFID technology provides an excellent backbone for the building of low-cost large-scale sensor networks (Hussain et al. 2009). Real-world implementation of RFID-based sensors to measure strain, temperature, water quality, and noxious gases has confirmed the feasibility of these types of sensors as low-cost wireless autonomous sensors, which ultimately enable pervasive cognitive networks for various IoT applications (including the agriculture domain) (Cook et al. 2014). Considering the type of sensing mechanism, these types of RFID sensors can be classified into the following categories (Rayhana et al. 2021).

RFID Tag Chip-based Sensors—the sensing unit forms an integrated part of the tag microchip and can operate at low, medium, high, and ultra high frequencies. These types of sensors have been used to measure different types of parameters i.e., water, moisture, temperature, gas, light intensity, etc.

Chipless RFID Tag-based Sensors—instead of using a microchip, an electromagnetic signature is used to encode sensed information that performs well in the presence of interference. These sensors have been used for the monitoring of moisture, humidity, temperature, crack, etc.

RFID Tag Chip with External Sensor—This category indicates the integration of standalone sensors with the microchip of RFID tags. Sensors that fall under this category can be used to measure temperature, moisture, humidity, strain, gas, viscosity, etc.

RFID Tag Antenna-based Sensors—these types of sensors can be fabricated on a low-cost substrate (i.e., plastic, fabric, paper) and do not require to be integrated with an external sensor. Mainly the working of these sensors is based on measuring the changes in RFID tags' antenna parameters i.e., resonance frequency, phase, gain, etc. to monitor various environmental factors i.e., moisture, temperature, gas, strain, etc.

Soil monitoring, plant growth monitoring, environmental monitoring, and harvest quality monitoring are prominent areas of agriculture where RFID-based sensor technology has been widely implemented.

3.6.1   Soil Monitoring with RFID-Based Sensors

The most essential soil parameters that ensure the healthy growth of crops/plants are moisture, salinity, and pH value (Vena et al. 2018). Different types of RFID-based sensors have been developed for soil monitoring; for example,

- To measure soil moisture, smart sensor nodes using 2.4 GHz active RFID tags are developed that are ultimately used at agriculture farms to optimize the usage of water (Hamrita and Hoffacker 2005).
- To measure soil salinity, chipless and chip-based (UHF) RFID sensors have been developed (Dey et al. 2016, 2019).
- A Chipless (UWB) RFID-based pH sensor is developed that can accurately calculate pH values between 4 and 7 (Athauda et al. 2020).

3.6.2   Plant Growth Monitoring with RFID-Based Sensors

Chip-based RFID sensors have been used to observe plant height, size, number of flowers/fruits/vegetables, etc. in an agricultural field. Moreover, image acquisition and image analysis through the use of embedded camera sensors show that the use of microchips is very helpful in measuring plant growth (Rayhana et al. 2021).

3.6.3   Environmental Monitoring with RFID-Based Sensors

The continuous and accurate monitoring of environmental parameters i.e., light intensity, temperature, humidity, oxygen level, etc. ensures the healthy growth of crops in an agricultural field or greenhouse. Two examples of state-of-the-art RFID-based sensor systems to measure environmental parameters have been briefly discussed below.

Fig. 3.3 Placement of temperature sensors on leaves

RFID-based Wireless Temperature Sensing System that is developed to measure the water stress level of plants (Palazzi et al. 2019). This work is proposed to implement precision agriculture with the development of a leaf-compatible wireless temperature sensing system with integrated RFID technology. For wireless connectivity, RFID transponders have been selected due to reduced installation/maintenance costs. The implemented scenario is shown in Fig. 3.3. Figure 3.3 describes that a few temperature sensors are placed on leaves (direct contact) and others in the air placed in the same environmental conditions. Based on the principle that correctly hydrated plants' temperature (T_{Px}) is lower than air temperature (T_{Air}), the system can schedule irrigation after detecting the leaf-air temperature gradient.

Harvest Quality Monitoring System—Harvest quality of agricultural products is very important for both sellers and consumers because of their short shelf lives. RFID-based sensor technology is helpful in monitoring various environmental parameters (i.e., temperature, humidity, specific gas concentration, etc.) upon which the fruits/vegetables' freshness is dependent (Le et al. 2016). One of the typical examples of RFID-based sensor technology is to monitor food products as shown in Fig. 3.4.

One of the important steps in the evolution of RFID-based sensing systems is its employment in the Wireless Identification and Sensing Platform (WISP) devices (Wireless Identification and Sensing Platform. https://sensor.cs.washington.edu/WISP.html). WISP devices can sense different physical quantities (i.e., temperature, light, liquid level, acceleration, etc.) along with the energy harvested through the received reader's signal. This WISP technology permits the creation of RFID sensor networks that ultimately require no batteries (Philipose et al. 2005).

3.7 Smart Phone Technology

It is observed that the emergence of low-cost smartphones significantly changes the real-life scenario of human lives. Interactiveness to digital systems through smartphones has become a cornerstone in almost all modern-day application developments.

Fig. 3.4 RFID-based
sensing systems to monitor
food quality

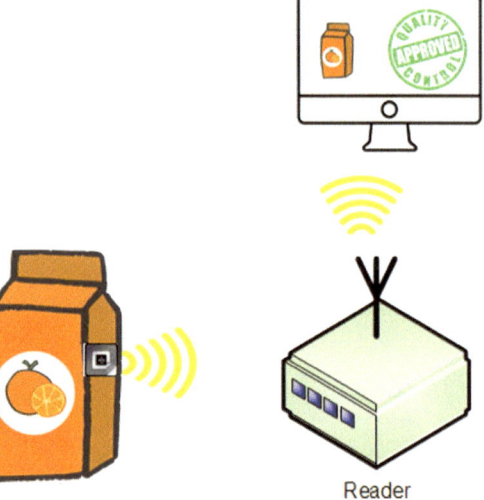

Reader

A large number of agricultural mobile apps have been developed to promote the real-ization of smart farming systems. These agricultural mobile apps can be categorized as (Farooq et al. 2019).

Soil Sampling Apps—assist farmers by submitting and viewing soil test information that ultimately helps in decision-making about recommended land treatments by proper fertigation.

Agriculture Information Management Apps—to help farmers in decision-making about farm activities i.e., planting, spraying, harvesting, etc.

Precision Agriculture Apps—allow remote monitoring, controlling, and precise scheduling of farm resources.

Weather Information Apps—to provide local weather forecasts and hourly weather alerts.

Agricultural Products Pricing Apps—provide detailed information about past and present market price changes about a certain agricultural product and help in controlling the productivity of that specific product by forecasting profit-making.

Agriculture Calculator Apps—to help farmers to calculate seed rate, pesticide dose, plant population, fertilizer blend rate, etc.

Table 3.4 Comparison of mobile remote sensing platforms

Platform	Observation area	Coverage area	Observation frequency	Field view	Ground resolution	Cost
Satellite	Worldwide	10 km	Day(s)	Narrow	5 m/pixel	High cost
Airborne	Regional	1 km	Hour(s)	Wider	0.5 m/pixel	High cost
UAV	Local	100 m	Minute(s)	Wide	0.05 m/pixel	Cost effective

3.8 Remote Sensing

Remote sensing is fundamentally related to the acquisition of an object's information from a distance without using any physical contact. In digital agriculture, static and mobile remote sensing has been widely implemented to capture field data. The static (also known as Ground-based) remote sensing technologies include the use of video cameras and microphones. Video cameras are utilized to monitor crops, livestock, and farm equipment. Microphones are usually utilized to capture sounds of farm pests, farm machinery for ongoing operations, and changes in animal vocalizations (indicating distress or illness).

On the other hand, the use of three mobile remote sensing platforms (Shafi et al. 2019; Matese et al. 2015) has been explained below:

- Satellite-based remote sensing platforms to capture high-resolution imagery and agricultural field data
- Airborne-based remote sensing platforms consisting of aircraft with sensing cameras used to obtain field images
- Unmanned Aerial Vehicle (UAV)-based sensing platforms consisting of fixed-wing drones having high-resolution cameras or multispectral sensors used for aerial imagery to identify areas of stress, field monitoring, soil and environmental monitoring, etc.

Characteristics comparison of all these mobile remote sensing platforms is shown in Table 3.4.

3.9 Development Boards and Data Acquisition

A Development Board (also known as a Microcontroller Development Board) is a hardware platform or printed circuit board with a mounted microcontroller/microprocessor on it and is intended to facilitate system designers to develop electronic systems and IoT projects. Key components (other than microcontroller/microprocessor) of development boards include memory, input/output interfaces, communication modules (Bluetooth, WiFi, Ethernet, etc.), and Integrated Development Environments (IDEs) or software development tools. There are two main types of Development Boards i.e., Single-Board Microcontroller (SBM) and Single-Board Computer (SBC).

3.9.1   Single-Board Microcontroller (SBM)

The SBM contains one or more processors, memory, and programmable input/output peripherals on a single integrated circuit. These SBMs are different from micro-processors (available in PCs) and are designed for embedded applications (Mathur 2020). The cost-effectiveness of these SBMs supports their usage for the addition and enhancement of computing capabilities of a smart thing. Below are the significant features of SBMs:

- SBMs have a specific amount of RAM
- SBMs have flash memory to store offline data
- SBMs have input/output pins to connect sensors/actuators
- SBMs have Ethernet/WiFi ports for Internet connectivity
- SBMs have power supply pins to supply power to attached sensors

The most popular SBM is Arduino, which is used in both small-scale (by novice learners in individual projects) and large-scale (by professionals in industry projects) settings to develop devices or systems that interact with the environment through sensors and actuators (Arduino. https://www.arduino.cc/). Figure 3.5 shows a general-purpose Arduino UNO SBM with different components. Table 3.5 presents several basic wireless nodes equipped with microcontrollers and sensors that are commonly used in the agricultural domain (Shafi et al. 2019).

3.9.2   Single-Board Computer (SBC)

The Single-Board Computer (SBC) is a handy and compact but completely func-tional computer, which is built on a single circuit board with all other peripherals (i.e., microprocessor, Random Access Memory (RAM), USB and HDMI ports, I/O functionality, Ethernet, and WiFi Sockets, etc.) on it. In simple words, it is a single circuit board that is capable of supporting all types of processing. Initially, these boards were built for educational purposes but later utilized in many mainstream industrial and commercial applications (Mathur 2020). Figure 3.6 shows a general-purpose (the most famous) Raspberry Pi SBC (Raspberry Pi. https://www.raspbe rrypi.org/) with different components.

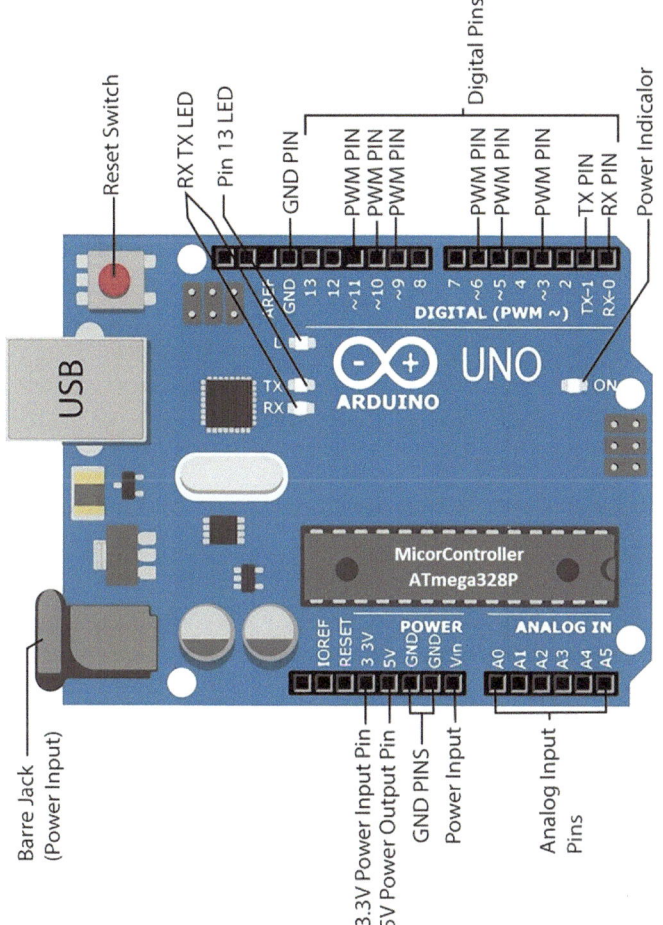

Fig. 3.5 Arduino UNO SBM with components

3.9.3 SBM and SBC Comparison

Both SBM and SBC are preferred over personal computers and server devices due to many advantages associated with them i.e.,

- small in size (almost palmtop size) and easily portable
- handy in use (with the support for different operating systems i.e., Linux, Windows IoT)
- power-efficient

Table 3.5 A few wireless nodes (with microcontrollers and sensors) used in the agriculture domain

Node	Microcontroller	Sensor attached
IRIS	ATmeaga128L	Temperature, humidity, barometric pressure, light
Cricket	ATmega128L	Temperature, humidity, light, barometric pressure, video sensor, GPS
MICAz	ATmega128L	Temperature, humidity, barometric pressure, light
MICA2DOT	ATmega128L	Temperature, humidity, light, barometric pressure, video sensor, GPS
MICA2	ATmega128L	Temperature, humidity, light, barometric pressure, video sensor, GPS
TelosB	TIMSP430	Temperature, humidity, light
Imote2	Marvell/XScalePXA271	Temperature, humidity, light

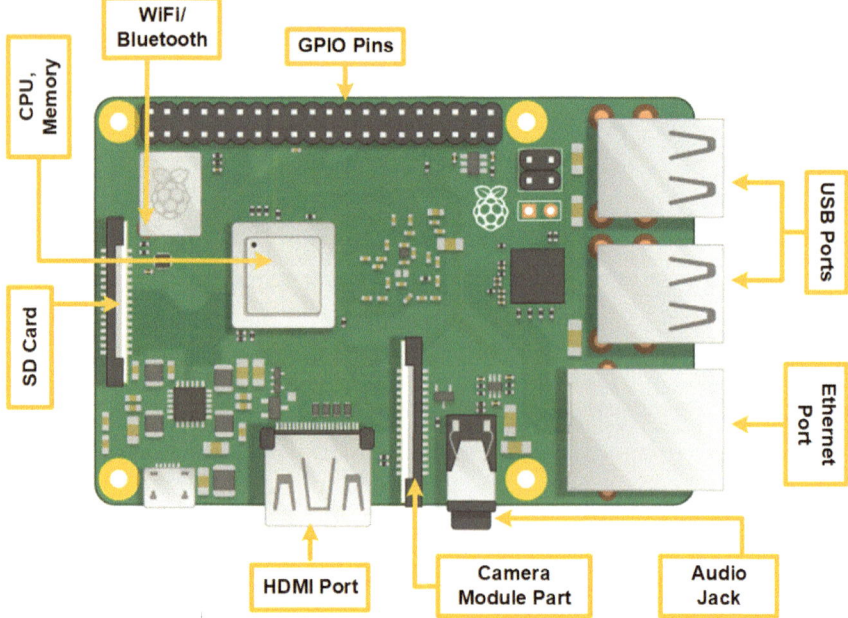

Fig. 3.6 Raspberry Pi SBC with components

- contain I/O pins and GPIO (General Purpose Input Output) pins, which are used to support the communication with auxiliary hardware i.e., sensors, actuators, LEDs, other MCUs, etc.
- useful for the prototyping of an IoT system and enable users to quickly connect sensors and actuators with these boards. Moreover, the accompanying software of these development boards facilitates the deployment of code for these sensor devices. There are several SBMs and SBCs available from different companies

i.e., Arduino, Raspberry Pi, Samsung, etc. Although, the selection of the right development board and MCU ultimately depends on the nature of the application; however, different factors also play an important role i.e.,

- easy availability and compatibility of development board to support sensors of your application
- sufficient memory to execute your IoT application
- energy-efficient architecture and cost of development board to implement IoT system.

3.9.4 Role of Development Boards in Agricultural Data Acquisition

SBMs and SBCs play a significant role in agricultural data acquisition with the provisioning of cost-effective platforms. The key roles involve the

- integration of a variety of agricultural sensors to collect field environmental data i.e., temperature, humidity, soil moisture, light sensors, etc.
- integration of camera, microphone, and RFID technology to monitor livestock health and behavior.
- automation of farm machinery and equipment.
- data logging over time to enable monitoring and recording of various environmental parameters to support proactive maintenance.
- usage in educational programs to teach students about agricultural data acquisition
- creation of cost-effective weather stations through integrating weather-related devices i.e., rain gauges and anemometer.
- Control and coordination of various agricultural equipment (based on GPS data) to facilitate farmers for precise planting, fertilizing, and harvesting.

Questions

Q3.1: The use of drones integrated with RFID-based sensing technology is one of the solutions that has been proposed to collect data from agricultural fields. Elaborate this concept and its benefits with reference to the scenario shown in Fig. 3.7.

Q3.2: The digitalization of aquaponic systems is possible through the deployment of cost-effective sensors to collect data from fish tanks (or aquariums) and grow beds. Describe the types of data and the types of sensors that are required to be collected and deployed respectively for aquaponic systems as shown in Fig. 3.8.

Q3.3: The architecture of the Digital Dairy Farming system is shown in Fig. 3.9. Explain the usage of digital technologies shown in Fig. 3.9 related to data acquisition at this dairy farm system. Also, elaborate the flow of data collected from sensors and digital devices (deployed and installed at different locations of this dairy farm system) to cloud storage.

Fig. 3.7 RFID-based drone
to collect data from plants
(Figure for Question 3.1)

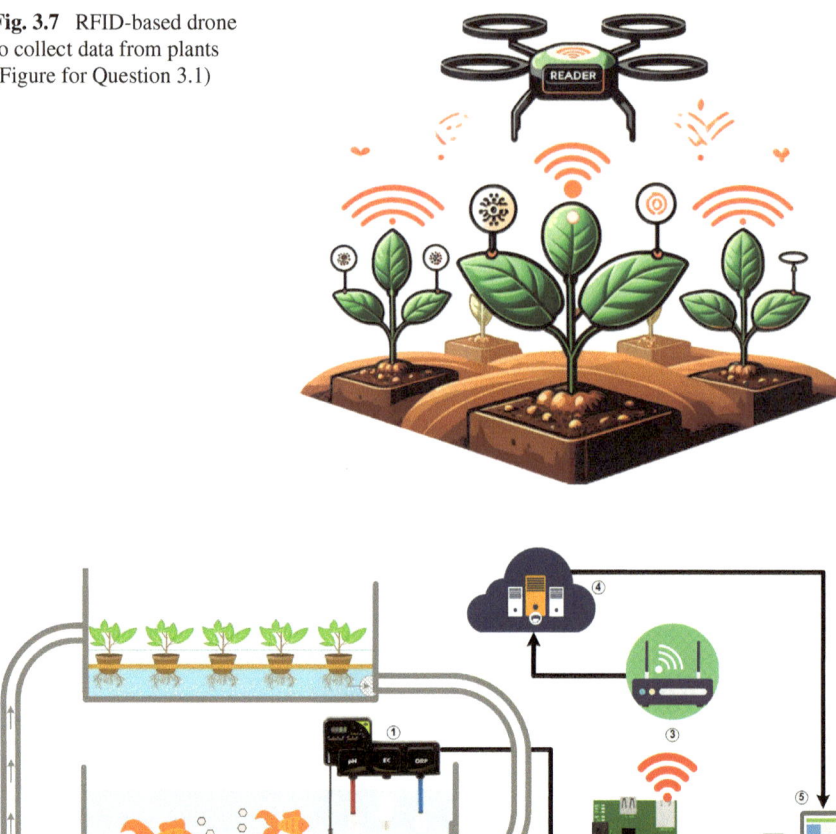

Fig. 3.8 IoT-based aquaponic system (Figure for Question 3.2)

Q3.4: Explain the use of video cameras and microphones to monitor livestock activities in an agricultural farm.

Q3.5: Low temperature, high humidity, low luminosity as well as low moisture and decrease of oxygen level in the soil cause a lower development of Plantain. Write the names of sensors and IoT devices that are required to be deployed to collect data for these parameters along with the growth of Plantains in an IoT-based solution that is implemented to handle this issue of lower Plantain development as shown in Fig. 3.10.

Fig. 3.9 Architecture of digital dairy farming system (Figure for Question 3.3)

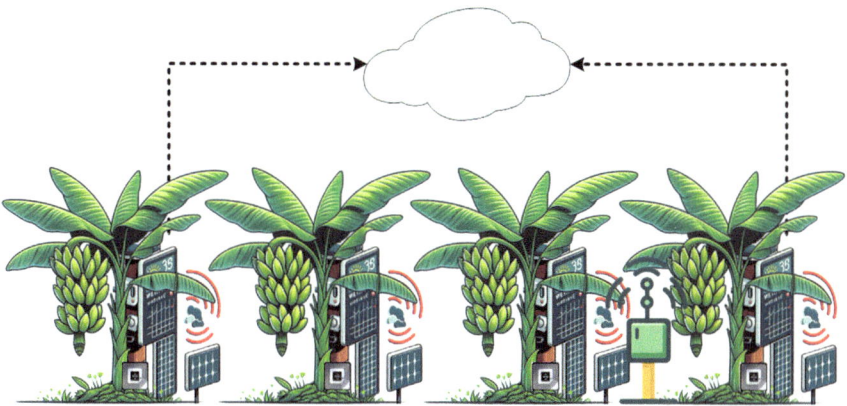

Fig. 3.10   Banana plantain (Figure for Question 3.5)

References

Arduino. https://www.arduino.cc/
Athauda T, Banerjee PC, Karmakar NC (2020) Microwave characterization of chitosan hydrogel and
    Its use as a wireless pH sensor in smart packaging applications. IEEE Sens J 20(16):8990–8996

Ayaz M et al (2019) Internet-of-things (IoT)-based smart agriculture: toward making the fields talk. IEEE Access 7:129551–129583

Cook BS et al (2014) RFID-based sensors for zero-power autonomous wireless sensor networks. IEEE Sens J 14(8):2419–2431

Dey S et al (2016) Electromagnetic characterization of soil moisture and salinity for UHF RFID applications in precision agriculture. In: Proceedings of the 46th IEEE European microwave conference (EuMC)

Dey S et al (2019) A folded monopole shaped novel soil moisture and salinity sensor for precision agriculture based chipless RFID applications. In: IEEE MTT-S international microwave and RF conference (IMARC)

Elijah O et al (2018) An overview of internet of things (IoT) and data analytics in agriculture: benefits and challenges. IEEE Internet Things J 5(5):3758–3773

Farooq MS et al (2019) A survey on the role of IoT in agriculture for the implementation of smart farming. IEEE Access 7:156237–156271

Fraden J (2004) Handbook of modern sensors: physics, designs, and applications. Springer

Hamrita TK, Hoffacker EC (2005) Development of a "smart" wireless soil monitoring sensor prototype using RFID technology. Appl Eng Agricult 21(1):139–143

Hunt VD, Puglia A, Puglia M (2007) RFID: a guide to radio frequency identification. Wiley

Hussain S, Schaffner S, Moseychuck D (2009) Applications of wireless sensor networks and RFID in a smart home environment. In: IEEE seventh annual communication networks and services research conference

Le GT et al (2016) Long-range batteryless RF sensor for monitoring the freshness of packaged vegetables. Sens Actuat A Phys 237:20–28

Luvisi A (2016) Electronic identification technology for agriculture, plant, and food: a review. Agron Sustain Develop 36(1):13

Matese A et al (2015) Intercomparison of UAV, aircraft and satellite remote sensing platforms for precision viticulture. Remote Sens 7(3):2971–2990

Mathur P (2020) IoT machine learning applications in telecom, energy, and agriculture

Miorandi D et al (2012) Internet of things: vision, applications and research challenges. Ad Hoc Netw 10(7):1497–1516

Ojha T, Misra S, Raghuwanshi NS (2015) Wireless sensor networks for agriculture: the state-of-the-art in practice and future challenges. Comput Electr Agricult 118:66–84

Palazzi V et al (2019) Leaf-compatible autonomous RFID-based wireless temperature sensors for precision agriculture. In: IEEE topical conference on wireless sensors and sensor networks (WiSNet)

Philipose M et al (2005) Battery-free wireless identification and sensing. IEEE Pervasive Comput 4(1):37–45

Rahaman MM, Azharuddin M (2022) Wireless sensor networks in agriculture through machine learning: a survey. Comput Electr Agricult 197:106928

Raj P, Raman AC (2017) The internet of things: enabling technologies, platforms, and use cases. Auerbach Publications

Raspberry Pi. https://www.raspberrypi.org/

Rayhana R, Xiao G, Liu Z (2021) Rfid sensing technologies for smart agriculture. IEEE Instrum Measur Mag 24(3):50–60

Ruiz-Garcia L et al (2009) A review of wireless sensor technologies and applications in agriculture and food industry: state of the art and current trends. Sensors 9(6):4728–4750

Shafi U et al (2019) Precision agriculture techniques and practices: from considerations to applications. Sensors 19(17):3796

Vena A et al (2018) An inkjet printed RFID-enabled humidity sensor on paper based on biopolymer. In: Proceedings of the 2nd IEEE URSI Atlantic radio science meeting (AT-RASC)

Wireless Identification and Sensing Platform. https://sensor.cs.washington.edu/WISP.html

Chapter 4
Data Communication in Digital Agriculture

4.1 Learning Objectives

After studying this chapter, students will be able to

- enlist the name of communication technologies that are part of the agricultural field data transmission model
- elaborate the role of Short-Range, Long-Range, Cellular, and Satellite communications in digital agriculture
- explain the impacts of Broadband technology on agricultural profitability

4.2 Agricultural Field Data Transmission

A typical IoT-based farm monitoring system involves the deployment of various sensors and digital devices at agricultural fields to collect soil, plant/livestock, and environmental data. Typically, the collected data is required to be transmitted to storage (cloud) servers for further processing and analysis. The data transmission from sensors and digital devices to storage servers is possible through various wireless (Bluetooth, Wi-Fi, Zigbee, etc.), cellular (2/3/4/5 G, LTE, LPWAN, etc.), satellite (Satellite Internet, GPS, GIS, etc.), and wired (i.e., Broadband) communication technologies. Therefore, communication technologies are an integral and important part of Digital Agriculture systems. The end-to-end data flow in a typical digital agriculture system may involve different communication technologies at different stages of data transfer. Mainly the transmission of agricultural data (vital field parameters i.e., soil, crops, livestock, weather, etc.) from field sensors/devices to edge routers or gateways is based on wireless communication technologies. However, data transfer from edge routers/gateways to large-scale data centers is primarily dependent on Broadband technology. Each communication technology (protocol) has specific pros and cons over others and it is not easy to select the most suitable communication technologies for a particular digital agriculture system (Avşar and Mowla 2022). Therefore, at the design stage, it is crucial to consider the characteristics and capabilities of available communication technologies to fulfill the application-specific requirements of the digital agriculture system. The characteristics, capabilities, and functionality perspectives of these wireless and wired communication technologies have been discussed below.

4.3 Wireless Communication

The collection of remote data from agricultural fields demands continuous connectivity that is possible through the availability of various heterogeneous wireless communication technologies. The selection of appropriate communication technology largely affects the overall efficiency of the system. Therefore, it is important to understand the connectivity spectrum and associated features of different wireless technologies. Based on their connectivity spectrum, these advanced and frontier wireless technologies (shown in Fig. 4.1) can be categorized as (Holdowsky et al. 2015; Lin et al. 2017).

- Short-Range Communication Technologies (i.e., RFID, NFC, Bluetooth, ZigBee, etc.)
- Long-Range Communication Technologies (i.e., LoRaWAN, SigFox, etc.)
- Wireless Local Area Communication Technologies (WiFi, Wi-Fi HaLow, etc.)
- Cellular Communication Technologies (2G/3G/4G/5G, LTE, NB-IoT)
- Satellite Communications

The comparison of wireless communication technologies in terms of range, data rate, and power consumption is shown in Fig. 4.2.

4.3.1 Short-Range Communication Technologies

Short-range communication technologies support local communication between devices that are in close proximity and intend to transfer data over short distances

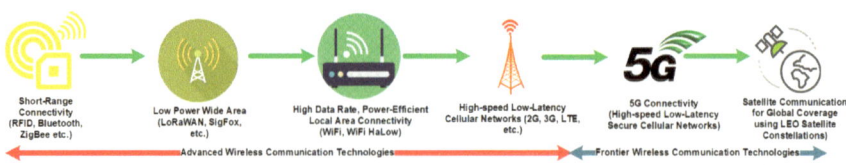

Fig. 4.1 Advanced and frontier wireless communication technologies

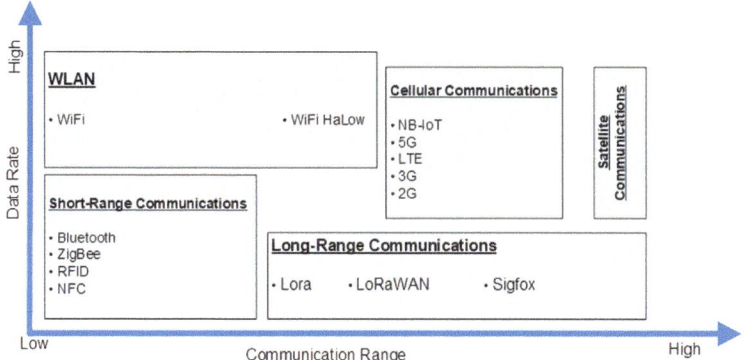

Fig. 4.2 Data rate and communication range comparison of wireless communication technologies

(from a few meters to a few hundred meters). RFID, NFC, Bluetooth, and ZigBee are prominent examples of short-range communication technologies.

RFID

RFID is one of the types of short-range communication technologies that use electromagnetic waves to transfer data between devices. The components and working details of the RFID system along with their usage in agriculture scenarios have already been discussed in Chap. 3 from a data acquisition perspective.

NFC

Based on RFID, NFC (stands for Near-Field Communication) is an integrated circuit that allows short-range wireless communication to exchange only small bits of information between two devices within close proximity of each other (over a distance of ≤10 cm). NFC communication systems consist of two parts i.e., a reader chip (active device) and tag (passive device) and in terms of working it operates in two modes i.e., active mode and passive mode. In active mode, both devices reader and tag generate their own radio frequency fields. In passive mode, the chip part generates a radio frequency field that is used to provide power, read information, and send NFC commands to NFC tags. From a usage perspective, NFC supports three modes of operations i.e., Read/Write mode (that allows NFC-enabled devices to read/write passive tags), Peer-to-Peer mode (that allows NFC devices to exchange data), Card Emulation mode (allows NFC device to act as NFC card that can be accessed by external NFC reader) (Coskun et al. 2013) as shown in Fig. 4.3.

In real-life scenarios, payments, transportation, access control, etc. through NFC-enabled phones and cards are typical examples of NFC technology. Similarly, NFC technology has several potential applications in digital agriculture (Wan et al. 2019; Ahiara 2022) i.e., farmers can use NFC-enables devices to

- read information from implanted NFC tags for animal identification, tracking animal health records, monitoring animal movement, vaccination schedules, etc.

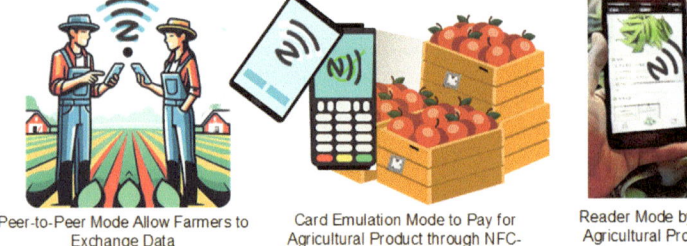

| Peer-to-Peer Mode Allow Farmers to Exchange Data | Card Emulation Mode to Pay for Agricultural Product through NFC-Enabled Mobile Phone | Reader Mode by Mobile Phone to Get Agricultural Product Information from Attached Tags |

Fig. 4.3 Examples of three modes of NFC operations

- track the movement of agricultural equipment/machinery and products i.e., harvested crops.
- trace information about the origin and production methods of agricultural products.
- check the status of irrigation equipment to optimize the usage of water in the field.
- buy and sell agricultural products through contactless payments.
- share data with other farmers and agricultural stakeholders e.g., pesticide and fertilizer companies.

Bluetooth

Bluetooth is also a wireless technology that supports short-range communication for exchanging data between devices (Bisdikian 2001). Currently, a version of Bluetooth known as Bluetooth Low Energy (or Bluetooth LE) (Liu et al. 2021) is preferred in the design and implementation of various digital systems due to the reason of minimal energy consumption. Like RFID and NFC, Bluetooth technology has several applications in agriculture (i.e., soil, environment, crop, and livestock monitoring, equipment connectivity with farm machinery, control of irrigation equipment, inventory, asset management in agriculture, etc.) (Singh and Sobti 2021; Drougka et al. 2013).

ZigBee

Zigbee wireless communication technology (Pan and Tseng 2007) stands out as a favored choice in the realm of digital agriculture system implementations due to its unique set of attributes that are aligned well with the demands of modern agricultural practices. Zigbee is well-regarded for its cost-effectiveness, low power consumption, reliable, and secure data transmission that make it ideal for its deployment in remote agricultural locations. Below are some key reasons describing how Zigbee is fostering efficient and resilient farming practices while accommodating various applications crucial in digital agriculture (Singh and Sobti 2021; Hidayat 2017).

- The cost-effectiveness of ZigBee makes it affordable for farmers for large-scale deployments in their agricultural lands.

- ZigBee characteristic of low power consumption extends the battery replacement periods and makes it suitable for battery-powered sensors/devices deployed in remote agricultural locations.
- ZigBee's support for mesh networking allows sensors to form a self-healing network that ultimately enables seamless communication (even in the presence of disruptions), reliability, and coverage in large agricultural areas.
- The versatility and adaptability of ZigBee make it suitable for a wide range of digital agriculture applications i.e., environmental sensing, precision irrigation, equipment control, livestock tracking, etc.
- Based on open standards, ZigBee promotes interoperability between devices of different manufacturers that supports the development of scalable digital agriculture systems.
- The short to medium-range communication of ZigBee contributes to better network management and is suitable for the distributed deployment of agricultural sensors across large-scale agricultural fields.
- ZigBee's data rate is sufficient to fulfill the demands of agricultural applications that do not require field sensors to transmit data at high data rates.
- The low-duty cycle ZigBee devices make them suitable for those agricultural applications where periodic information update is required i.e., irrigation management, water quality management, pesticide, and fertilizer control systems.

Table 4.1 provides a comparison of RFID, NFC, Bluetooth, and ZigBee on associated features and characteristics.

Table 4.1 Comparison of RFID, NFC, bluetooth, and ZigBee

Characteristic	RFID	NFC	Bluetooth	ZigBee
Communication range support	Short to long	Very short	Short to medium (Bluetooth5 supports long-range)	Mostly short to medium (some variants support long-range communication)
Power requirements	Passive (powered by reader, without battery)	Passive (powered by reader, without battery)	Active and passive	Low power (battery operated)
Device interaction requirements	Requires proximity (No direct interaction)	Requires close proximity (touch-based)	Requires pairing, typically longer range	Requires network formation, mesh networking
Data transfer	Rate varies (low to high)	Up to 424 kbps	1–3 Mbps	20–250 kbps

4.3.2 Long-Range Communication Technologies

Long-range communication technologies are designed to transmit data over long distances (kilometers and miles) with and without proper infrastructure. Some notable long-range communication technologies (i.e., LoRaWAN and Sigfox) that play a crucial role in digital agriculture have been discussed below.

LoRa and LoRaWAN

LoRa (Long-Range) (Sundaram et al. 2019) and LoRaWAN (Long-Range Wide Area Network) (de Carvalho Silva 2017) communication techniques are often used in combination to support low-power and long-range wireless communication in the context of IoT. Essentially, LoRaWAN based on LoRa technology defines the system architecture for the IoT network and enables secure communication between digital devices and the network server. Primarily, LoRa is used in scenarios where digital devices need to directly communicate with each other over long distances without proper network infrastructure (known as point-to-point communication). However, LoRaWAN supports end-device communication with network servers (through gateways) in a proper network infrastructure established with gateways and network servers as shown in Fig. 4.4. From Fig. 4.4, it becomes evident that LoRaWAN architecture consists of two parts i.e., frontend (consisting of end nodes and gateways) and backend (consisting of network servers).

Bidirectional communication, data aggregation, and device management are the main features of LoRaWAN. LoRaWAN supports a variety of digital agriculture applications including farm monitoring, livestock monitoring, irrigation control, weather station deployments, etc. (Miles et al. 2020). Below are some key reasons for preferring LoRaWAN in digital agriculture.

- LoRaWAN's long-range communication is suitable for the remote nature of agricultural environments as it allows sensors and IoT devices to communicate over long distances.
- LoRaWAN's characteristic of low power consumption supports the deployment of battery-powered sensors in agricultural fields as it ensures device operability over an extended period of time without frequent battery replacements.

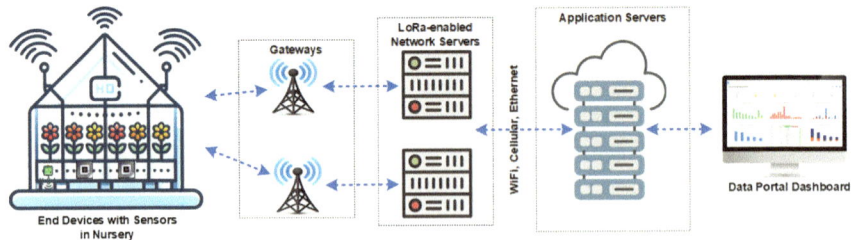

Fig. 4.4 LoRaWAN architecture with frontend gateways and backend servers

- LoRaWAN's capability of transmitting small amounts of data at a low rate saves device energy (and prolongs the battery life of devices) that are deployed in agricultural fields.
- LoRaWAN's ability of resilience to interference makes it appropriate to be deployed in harsh environments of outdoor agricultural fields.
- LoRaWAN's simple and easy deployment allows agricultural stakeholders to set up sensor networks quickly which is crucial in dynamic agricultural environments.
- LoRaWAN's support for interoperability between devices from different manufacturers allows agriculturists to go for the best solutions in terms of selecting suitable sensors and devices to be deployed in agricultural fields to fulfill their specific needs.
- LoRaWAN's robust security ensures the integrity and confidentiality of sensitive agriculture information i.e., crop conditions and livestock health.

Sigfox

Sigfox is another low-power, long-range, and wide-area wireless communication technology that is designed for IoT applications (Fourtet and Ponsard 2020). Sigfox's architecture is centralized where a single base station can send data to Sigfox cloud. Sigfox sensors (end devices) directly send data to the Sigfox cloud and there is no need for gateways in Sigfox architecture as shown in Fig. 4.5. This simplified network management approach of Sigfox does not support its suitability for large-scale deployments and does not offer more flexibility to customize applications.

Like LoRaWAN, Sigfox's long-range, energy-efficient, and cost-effective data transmission capabilities support its use in digital agriculture applications i.e., remote crop/livestock monitoring, asset tracking, weather station deployments, etc. and ultimately enhance yield and sustainability in agriculture. Unlike LoRaWAN, the limited uplink/downlink data rates of Sigfox make it suitable for low-bandwidth agricultural applications such as sensor monitoring and livestock tracking. Both LoRaWAN and Sigfox have their own strengths and weaknesses and the choice between these two technologies for a specific application depends on various factors i.e., applications' data rate requirements, geographical coverage requirements, power consumption, network ownership preferences, initial deployment cost, etc. (Islam et al. 2020). A

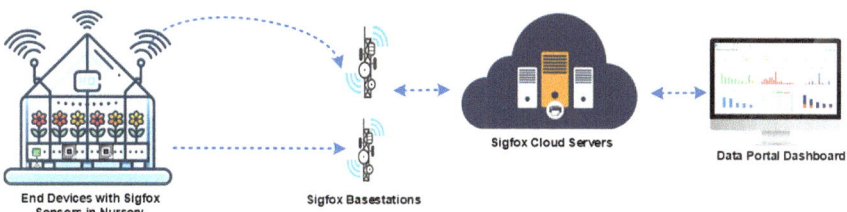

Fig. 4.5 Sigfox architecture support direct data transfer from Sigfox end devices to Sigfox Cloud via Sigfox network

few scenarios of digital agriculture where Sigfox can be preferred over LoRaWAN have been discussed below.

- If digital agriculture solution needs to be deployed globally without complex network infrastructure. In these scenarios, Sigfox's easy deployment model makes it an attractive choice.
- If digital agriculture applications demand infrequent transmission of small amounts of data at a low rate.
- If agricultural fields are in remote rural areas where cellular coverage is scarce and sensor deployment spans over a wide area.
- If sensors and IoT devices in digital agriculture applications are required to be operated for long periods without frequent maintenance and battery replacement.

4.3.3 Wireless Local Area Communication Technologies

WiFi

WiFi technology commonly used for local area networking of devices harnesses connectivity over Internet access to optimize various aspects of agricultural operations by allowing IoT devices to exchange data by radio waves (Omar et al. 2016). Compatibility with other WiFi-enabled technologies and economic feasibility in terms of its operational and maintenance cost are the main reasons for its preference for smart farming systems over cellular and long-range communication technologies. WiFi-enabled sensors, IoT devices, and drones help agriculturists make decisions about planting, irrigation scheduling, fertilizer application, and pest control strategies by gathering real-time data on soil conditions, crop health, and other environmental factors. In addition, WiFi can revolutionize agriculture farms by ensuring ubiquitous connectivity at lower cost for various devices and farm machinery i.e., security cameras, GPS-enabled tracking collars to monitor livestock activities, driverless tractors for automatic spraying, robots for weeding, drones to monitor crop health, pest infestation, disease attacks, etc.

Wi-Fi HaLow

Wi-Fi HaLow (an extension of traditional WiFi technology) is another wireless communication standard that is exclusively designed to address the needs of low-power and long-range IoT applications (Tian et al. 2021). The preference for Wi-Fi HaLow over LoRaWAN and Sigfox in certain digital agriculture scenarios depends on the specific requirements of the application (Enriko and Gustiyana 2024). For example,

- If typical Wi-Fi infrastructure is already available on the farm. In these scenarios, the simple and straightforward deployment of Wi-Fi HaLow offers seamless integration with existing Wi-Fi.
- If digital agriculture application demands high data transmission at a faster rate.

- If there is a demand for a moderate range of device connectivity especially in smaller or medium-sized agricultural settings.
- It is suitable for real-time monitoring applications that need frequent and bursty data traffic.
- If reliable and sufficient energy resources are available for devices because Wi-Fi HaLow consumes more energy than LoRaWAN and Sigfox.

4.3.4   Cellular Communications

Cellular communications play a pivotal role in revolutionizing the way modern farms are managed and pave the way for digital farming techniques through the provisioning of reliable and enhanced connectivity. This integration of cellular technologies in agriculture enables agriculturists to deploy a multitude of sensors and digital devices in agricultural fields for real-time monitoring and continuous gathering of field data i.e., soil moisture, crop health, temperature, humidity, weather conditions, etc. This real-time monitoring and information availability allows agriculturists to make optimal decisions regarding irrigation, pesticide spray, fertilizer application, etc. Recently, Long-Term Evolution (LTE), 5G, and NarrowBand IoT (NB-IoT) are cellular communication technologies (Tang et al. 2021; Raja et al. 2024; Valecce et al. 2020) that play significant roles in the realm of digital agriculture. Based on key characteristics, Table 4.2 provides a comparison of these technologies.

Table 4.2 Key characteristics of cellular technologies used in digital agriculture

Characteristics	LTE	5G	NB-IoT
Communication range	Medium to long	Medium to long	Long
Data rate	High (Mbps to Gbps)	Very-high (Gbps)	Low to moderate (kbps to Mbps)
Energy efficiency	Moderate	Improved	High
Latency	Low (10–10 ms)	Ultra-Low (1 ms or lower)	Low (10 ms to several s)
Coverage	Widespread	Expanding	Excellent in challenging environments
security	Strong	Enhanced	Tailored for IoT
Applications support	Mobile broadband, IoT	Enhanced mobile services, IoT	IoT with low data needs

4.3.5 *Satellite Communications*

Satellite communication technology plays an important role in digital agriculture as the high-resolution imagery taken by satellites not only helps in monitoring large-scale agriculture fields but also provides insight into crop health, growth patterns, pest, disease, and weed detection, water and nutrient availability in agricultural land, and livestock monitoring (Inoue 2020). Concerning risk assessment, satellites by providing real-time weather data help agriculturists in anticipating and preparing for floods and storms. Moreover, the satellite-based Global Positioning System (GPS) (Hofmann-Wellenhof et al. 2001) and Global Information System (GIS) (Biehl 2007) support digital agriculture in various ways. Table 4.3 describes a few common use cases of GPS and/or GIS technologies in agriculture (Neményi et al. 2003; Sood et al. 2015).

4.4 Broadband Networks

The effective working of a digital agricultural system in rural areas demands the use of high-speed Internet access (to boost farm profitability) which is not possible without broadband networks. The Federal Communications Commission has defined Broadband as high-speed Internet access where high speed means data transmission within 200 Kbits/s to 30 Mbits/s (Rural telecom educational series 2012). Broadband provides the Internet (and Internet-related services) at a higher speed than commonly available dial-up services.

The two broadband delivery mechanisms include Wired Broadband and Wireless Broadband. Wired Broadband can be further divided into three types i.e., Digital Subscriber Line (DSL) provided by telephone companies over copper lines, Cable Modem provided by cable operators over coaxial cable, and Fiber provided by Telecommunications providers over fiber optic cable. On the other hand, Wireless Broadband is either provided via radio signals (Wi-Fi or Mobile service) through an antenna or by Satellite systems. The importance of the high-speed Internet is evident from the fact that in many areas of the world, the agriculture industry has the potential to flourish by adopting technological advancements, but the main barrier is the availability of the Internet. Moreover, high-speed Internet access can affect agricultural profitability in several ways (Griffith et al. 2013) i.e., by

- provisioning of real-time information about crop production, weather forecasting, market trends, and decisions
- automation of farm operations
- providing access to the global marketplace
- supporting human labor regarding the increase of planting and harvesting speed
- allowing farmers to use software for soil mapping, yield mapping, and guidance systems that require the broadband network to provide real-time accurate data about the water and fertilizer needs of crops in the agriculture field

Table 4.3 Common agricultural use cases of GPS/GIS systems

GPS/GIS use cases	Description
Farm planning	GPS/GIS systems by providing information about field area size, soil characteristics, crop characteristics, etc. help farmers for the planning of farmlands
Field mapping	Provide exact estimation of the field to identify areas for farming and non-farming activities
Soil sampling	GPS helps in determining the soil variability by the provisioning of necessary data
Soil mapping	GPS and GIS technologies are cheaper and save time to create survey maps of soil positions and locations
Accurate planting	Helps farmers with the planting of specific crops with the provisioning of information about spacing and depth which specific crop requires
Planting ratio determination	GPS helps planting or cultivation of specific crops that demand specific sowing distance by determining the ratio of this type of planting
Weed location	Weeds in a particular piece of farmland hinder the growth of a crop. GPS helps to find the weed patches in vast areas of farmlands
Autonomous vehicle guidance	GPS can accurately locate the position of a moving vehicle and is used in guidance systems of autonomous vehicles for various agricultural phenomena. For example, used for guidance of plow tractors, harvesters, seeders, etc.
Yield mapping	Helps to map out the yield expected from farmland based on information about land/crop characteristics
Support in harsh weather conditions	GPS helps farmers to work in bad weather conditions (especially during low visibility field conditions i.e., rain, fog, dust, etc.)
Crop harvesting	Determining field area where the crop is ready for harvesting
Crop scouting	Concerning the scouting of crops, GPS helps experts to provide the exact mapping of a particular area suitable for a particular crop In crop scouting, many factors (weeds, pests, insect infestations) about a specific piece of land are taken into account for decision-making about the cultivation of a particular crop in that area
Farm machinery tracking and handling	Helps remote tracking of farm machinery in farmland and can also assist with the direction of work
Irrigation management	Assist farmer with the provisioning of the following information i.e., • availability of water (or water resources) at a particular piece of farmland • about the irrigated and non-irrigated land area on a farmland • Identification of waterlogged areas on a farmland
Meteorological mapping	Assist in determining suitable crop types for a particular piece of farmland by mapping out certain climatic conditions
Personnel mapping	Helps in finding the presence, tracking, and mapping out of farmland personnel to measure their productivity in the agriculture farm

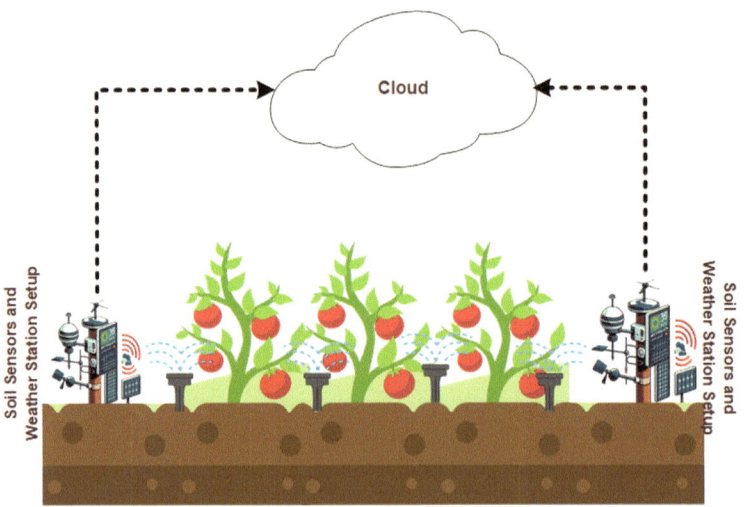

Fig. 4.6 Smart tomato orchard (Figure for Question 4.3)

- supporting software updates necessary to operate farm machinery and equipment

Questions

Q4.1: What is meant by Point-to-Point (P2P) communication of LoRa? Explain with the help of an example from an agricultural scenario.

Q4.2: What is meant by the low-duty cycle and how low-duty cycle of ZigBee devices make them suitable for agricultural applications?

Q4.3: For the quality production of Tomato fruits, farmers must maintain and monitor irrigation activities. To implement a smart irrigation system for a Tomato orchard, wireless sensors are required to be installed for the monitoring of soil water status in the Tomato orchard. These sensors transmit soil water level information to the cloud. Explain which types of wireless communication technologies can be used in this scenario of a large-scale Tomato field (as shown in Fig. 4.6).

Q4.4: The architecture diagram of the IoT-based Maize crop system is shown in Fig. 4.7. Illustrate why the system designer proposed the use of LoRaWAN for the development of this system.

Q4.5: The architecture diagram of the IoT-based Smart Flower Nursery is shown in Fig. 4.8. Explain the reason for selecting various communication technologies (as shown in Fig. 4.8) by the system designer to implement this smart flower nursery installation.

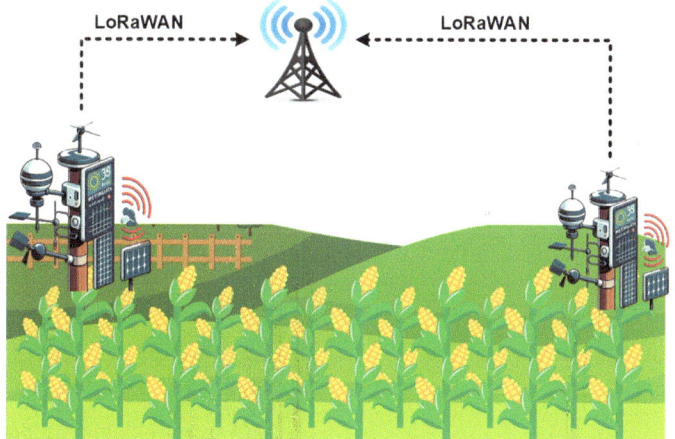

Fig. 4.7 Smart maize field (Figure for Question 4.4)

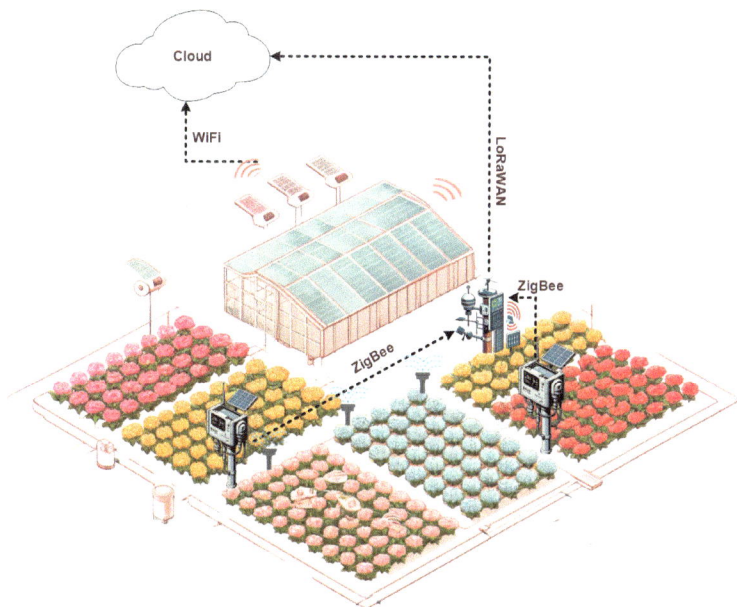

Fig. 4.8 Flower nursery (Figure for Question 4.5)

References

Ahiara WC et al (2022) Near field communication intelligent remote livestock monitoring system (Nigeria). J Sci Technol Res 4(2)

Avşar E, Mowla MN (2022) Wireless communication protocols in smart agriculture: a review on applications, challenges and future trends. Ad Hoc Netw 136:102982

Biehl M (2007) Success factors for implementing global information systems. Commun ACM 50(1):52–58

Bisdikian C (2001) An overview of the Bluetooth wireless technology. IEEE Commun Mag 39(12):86–94

Coskun V, Ozdenizci B, Ok K (2013) A survey on near field communication (NFC) technology. Wireless Personal Commun 71:2259–2294

de Carvalho Silva J et al (2017) LoRaWAN—a low power WAN protocol for internet of things: a review and opportunities. In: Proceedings of the 2017 2nd international multidisciplinary conference on computer and energy science (SpliTech)

Drougka M, Pontikakos C, Tsiligiridis T (2013) Bluetooth design configurations to support agricultural applications. Agricult Univ Athens Lab Inform Ieraodos 75(118):55

Enriko IKA, Gustiyana FN (2024) Wi-Fi HaLow: literature review about potential use of technology in agriculture and smart cities in Indonesia. In: Proceedings of the 2024 international conference on green energy, computing and sustainable technology (GECOST)

Fourtet C, Ponsard B (2020) An introduction to Sigfox radio system. LPWAN technologies for IoT and M2M applications. Elsevier, Amsterdam, pp 103–118

Griffith C et al (2013) Smart farming: leveraging the impact of broadband and the digital economy. CSIRO and University of New England, New England

Hidayat T (2017) Internet of things smart agriculture on zigbee: a systematic review. Jurnal Telekomunikasi Dan Komputer 8(1):75–86

Hofmann-Wellenhof B, Lichtenegger H, Collins J (2001) Global positioning system (GPS), in theory and practice. Springer, New York

Holdowsky J et al (2015) Inside the internet of things (IoT). Deloitte Insights

Inoue Y (2020) Satellite-and drone-based remote sensing of crops and soils for smart farming: a review. Soil Sci Plant Nutr 66(6):798–810

Islam N, Ray B, Pasandideh F (2020) IoT based smart farming: are the LPWAN technologies suitable for remote communication? In: Proceedings of the 2020 IEEE international conference on smart internet of things (SmartIoT)

Lin J et al (2017) A survey on internet of things: architecture, enabling technologies, security and privacy, and applications. IEEE Internet Things J 4(5):1125–1142

Liu C, Zhang Y, Zhou H (2021) A comprehensive study of bluetooth low energy. In: Journal of physics: conference series. IOP Publishing

Miles B et al (2020) A study of LoRaWAN protocol performance for IoT applications in smart agriculture. Comput Commun 164:148–157

Neményi M et al (2003) The role of GIS and GPS in precision farming. Comput Electr Agricult 40(1–3):45–55

Omar HA et al (2016) A survey on high efficiency wireless local area networks: next generation WiFi. IEEE Commun Surv Tutor 18(4):2315–2344

Pan M-S, Tseng Y-C (2007) ZigBee and their applications, in Sensor networks and configuration: fundamentals, standards, platforms, and applications. Springer, New York

Raja S, Subashini B, Prabu RS (2024) 5G technology in smart farming and its applications. Intelligent robots and drones for precision agriculture. Springer, New York, pp 241–264

Singh DK, Sobti R (2021) Wireless communication technologies for Internet of Things and precision agriculture: a review. In: Proceedings of the 2021 6th international conference on signal processing, computing and control (ISPCC)

Smart Agriculture and the Role of Broadband (2012) Rural telecom educational series. https://cco fkansas.com/resources/Documents/smart_agriculture.pdf

Sood K et al (2015) Application of GIS in precision agriculture. In: Paper presented as lead lecture in national seminar on precision farming technologies for high Himalayas

Sundaram JPS, Du W, Zhao Z (2019) A survey on lora networking: research problems, current solutions, and open issues. IEEE Commun Surv Tutor 22(1):371–388

Tang Y et al (2021) A survey on the 5G network and its impact on agriculture: challenges and opportunities. Comput Electr Agricult 180:105895

Tian L et al (2021) Wi-Fi HaLow for the internet of things: an up-to-date survey on IEEE 802.11 ah research. J Netw Comput Appl 182: 103036.

Valecce G et al (2020) NB-IoT for smart agriculture: experiments from the field. In: Proceedings of the 2020 7th international conference on control, decision and information technologies (CoDIT)

Wan X-F et al (2019) Near field communication-based agricultural management service systems for family farms. Sensors 19(20):4406

Chapter 5
Data Storage in Digital Agriculture

5.1 Learning Objectives

After studying this chapter, students will be able to

- describe the main enabling technologies for the storage of agricultural data
- elaborate the layered cloud service model for smart farming
- explains agricultural use cases where the coordination of edge/fog computing with cloud computing technology helps farmers for agricultural sustainability.

5.2 BigData and Agricultural BigData

Agricultural data storage is one of the critical components of digital agriculture that enables agriculturists to access vast amounts of heterogeneous data generated from various types of digital devices deployed in agricultural (indoor/outdoor) fields. In computing, this type of extremely large and diverse collection of data that continues to grow exponentially over time is known as BigData (Sagiroglu and Sinanc 2013). By definition, BigData refers to the large amount of organized, semi-organized, and unorganized data generated continuously at high speed and in large volume. Therefore, the key features of BigData are mainly summarized as three Vs i.e., volume (size of collected data from various sources), velocity (data collection at different speeds), and variety (data collection in diverse forms and different formats). In the realm of digital agriculture, these three Vs represent the huge volume of heterogeneous data encompassing data on soil, crops, farm machinery/equipment, pests, diseases, fertilizer, irrigation, animal behaviour, animal welfare, animal nutrition, animal acoustics, plant/animal growth videos, climate/weather, rainfall pattern, productivity or yield, aerial imagery, agri-food supply chain, agricultural stakeholders, etc. that is generated in different formats at different rates by various heterogeneous digital devices

M. A. Iqbal, *Digital Agriculture*,
SpringerBriefs in Agriculture, https://doi.org/10.1007/978-3-031-67679-6_5

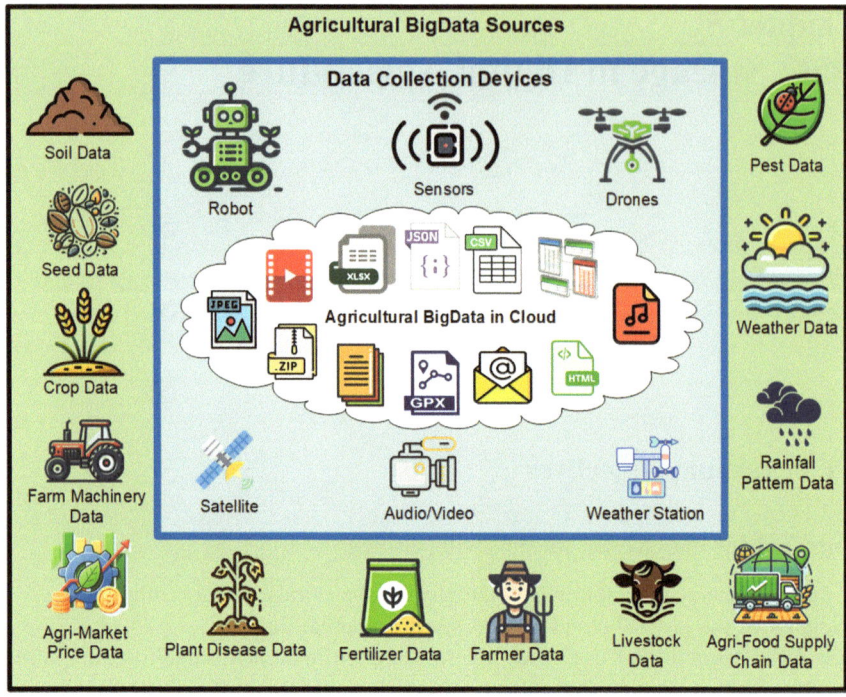

Fig. 5.1 Heterogeneous formats of agricultural data collected from heterogeneous agricultural sources using advanced digital devices

i.e., sensors, robots, weather stations, drones, satellites, microphone, (video) camera, etc. as shown in Fig. 5.1.

5.3   Driving Factors and Challenges of BigData Implication in Digital Agriculture

There are several pull factors and push factors that have been considered as the driving factors for BigData implication in agriculture (Wolfert et al. 2017) i.e.,

Pull factors—the factors that provide positive attraction to agricultural stakeholders through incentives and opportunities regarding the inclusion of advanced technologies to achieve new goals i.e., demand for more/better information, food safety, increased efficiency at low cost, better support/control management, improved decision-making, etc.

Push factors—the factors that compel or encourage agricultural stakeholders to achieve higher goals i.e., increased complexity of agricultural systems, need for prompt and enhanced decision-making, global security concerns, competitive pressures, etc.

Tables 5.1 and 5.2 provide an overview of the pull and push (driving) factors of BigData in digital agriculture (Cravero et al. 2022).

Although pull and push factors have contributed to the adoption of BigData in agriculture; however, BigData storage is a critical component and there are a few issues related to the implication of BigData in agriculture (Posadas and Gilbert 2020) (Carbonell 2016) i.e.,

Infrastructure and Connectivity—inadequate infrastructure and limited access to broadband networks hinder the real-time processing of BigData in developing countries.

Data Ownership and Dissemination—farm growers are concerned about data ownership and control as there are no obligations on agricultural technology providers to make data available only to concerned agriculturists. Farmers lose control over data and may have concerns about beneficiaries of derived data insights.

Table 5.1 Pull factors for BigData implication in digital agriculture

Pull factor	Description
Advancements in data-driven agriculture	Digital agriculture enabled by BigData attracts agriculturists to achieve efficient resource management and better decision-making
Increased access to advanced technology	Agriculturists are attracted to BigData because digital technologies become more accessible and ultimately help them to increase productivity with data-driven insights
Cost-effectiveness	Increase in overall efficiency at low cost regarding the monitoring and management of agricultural systems
Market demand for sustainable agriculture	Agriculturists are inclined to adopt BigData solutions to employ sustainable practices aligned with environmental considerations
Demand for better information	Better decision-making demands the optimized allocation of resources (water, pesticide, fertilizer) that is ultimately dependent on in-time better quality information
Increased profitability	Optimal decision-making and enhanced farm management based on digital data availability improve overall profitability
Optimized resource allocation	The use of BigData helps agriculturists in saving costs with the optimized utilization of farm resources
Food safety and nutrition security	Concerning their health and well-being, consumers are becoming more aware and concerned about food safety and nutrition security

Table 5.2 Push factors for BigData implication in digital agriculture

Push factor	Description
Challenges in traditional farming practices	Resource inefficiency, integrated pest management, and yield variability push agriculturists to adopt BigData solutions
Complexity of agricultural systems	The growing demands regarding the handling and analysis of large datasets to derive meaningful insight into modern digital agricultural systems ultimately urge agriculturists to adopt BigData solutions
Technological advancements	The availability of data collection from advanced digital technologies i.e., sensors, remote sensing, robotics, and advanced IoT devices with high computational power improves the decision-making process and pushes agricultural stakeholders to explore innovative solutions with BigData applications
Competitive pressure in the agricultural industry	To stay competitive, agricultural stakeholders are pushed to adopt BigData solutions offering cost savings and efficiency gains
Need for enhanced decision-making	To improve the decision-making process, there is a push to adopt BigData solutions that support actionable insights

Cost and Affordability—it is challenging for small-scale farmers with limited resources to bear the significant upfront costs of implementing BigData solutions.

Data Privacy—there are no rules regarding the privacy of data (collected by the agricultural technology providers) and therefore no guarantee that collected data will remain private. To encourage the use of digital agriculture technologies, the policies and rules are required to be drafted clearly.

Privacy and Security Principles—to encourage the use of digital agriculture technologies, the policies and rules are required to be drafted clearly to restrict unauthorized access or misuse of agricultural data.

Education and Ethical Considerations—proper education focusing on the implications of the right technologies is required as it can minimize the negative effects of BigData in agriculture.

Difficult Integration and Collaboration—lack of data format standardization restricts seamless collaboration and data sharing across different platforms.

Digital Literacy Gap—Agriculturists mostly lack the necessary skills to fully harness the potential of BigData in agriculture.

5.4 Types of Agricultural BigData

Considering the definition of BigData (mentioned in Sect. 5.1), BigData technology supports the storage of huge volumes of data that can be categorized as Structured (organized), Semi-structured (semi-organized), and Unstructured (unorganized) Data. The characteristics of all three data types along with relevant examples from the agriculture context have been discussed below.

5.4.1 Structured Data

There are two types of structured data i.e., Human-readable Structured Data and Machine-readable Structured Data.

Human-readable Structured Data refers to the formatted and well-organized data that is typically stored in structured file formats (mostly relational databases). The key characteristics of human-readable structured data include

- predefined schema that ensures uniformity and consistency in data storage
- organized data storage in tabular format with columns (representing data attributes) and rows (representing records)
- data element (stored in each cell of a table) is of a predefined data type i.e., integer, float, string, Boolean, date, time, etc.
- specific subsets of data can be retrieved (to generate reporting) through the use of Structured Query Language (SQL).

Below are some examples of structured data used in digital agriculture.

Soil Data: Data on soil properties in agricultural land i.e., soil type, soil texture, soil moisture, nutrient content, soil pH level, and compaction that assists agriculturists in nutrient management.

Crop Information: cultivated crops' details (i.e., crop type, crop variety, planting time, growth stages, obtained yield, etc.) that help agriculturists to plan optimum planting and harvesting time, crop rotations, etc.

Weather Information: meteorological data (i.e., temperature, wind speed, rainfall, humidity, etc.) of agricultural land that is crucial to predict weather patterns, monitoring environmental conditions, and decision-making about irrigation schedules.

Farm Equipment Data: data containing details about farm machinery/equipment, fuel consumption, maintenance records, etc. that is helpful in predictive maintenance.

Crop Pest and Disease Information: details regarding pest and disease incidence, pest lifecycle stages, and disease severity ratings required to support integrated pest management (IPM) strategies in early crop disease control.

Farm Financial Records: data containing information about farm expenses, profit, and revenue that is helpful for farmers/ranchers to monitor financial performance and plan farm budgets.

To store these types of data, the most used structured databases are relational databases that include Oracle, SQL Server, MySQL, PostgreSQL, etc. An example of structured data (Tables and Relationship between these Tables) stored in a relational database is shown in Fig. 5.2.

Machine-readable Interlinked Structured Data (also known as Linked Data) is built upon standard web technologies i.e., Uniform Resource Indicators (URIs), HyperText Transfer Protocols (HTTP), eXtensible Markup Language (XML), Resource Description Framework (RDF) (Miller 1998; Pan 2009), etc. The fundamental benefit associated with linked data is that it paves the way for the development of the Semantic Web. Semantic Web with the goal of enabling Internet data as machine-readable RDF and Web Ontology Language (OWL) (McGuinness and Harmelen 2004; Antoniou and Harmelen 2004) describe the conceptual relationships between entities on the World Wide Web (WWW). The creation of the semantic web ultimately enables computing machines to analyze all the data available on the web.

Concerning agricultural processes, there exist dependencies e.g., crop yield is dependent on land preparation, soil composition, climate conditions, insect/pest attack, etc. Therefore, different knowledge bases are required to store this kind of agricultural data. The semantic web helps to align these different isolated knowledge bases to allow these large datasets to be queried as a single interlinked resource for automatic analysis. However, the implementation of semantic web technology is directly reliant on the availability of existing linked data or semantic web sources. A few examples of semantic resources are

- Controlled vocabularies (set of preselected terms or words for a specific domain)
- Taxonomies (systematic arrangements of controlled vocabularies into the hierarchical structure)
- Thesauri (extension of hierarchical taxonomies with non-hierarchical relationships)
- Ontologies (formal structure of system).

Semantic resources for agriculture include the resources that use semantic technologies to describe the stored knowledge of agricultural processes by different individuals and organizations. The most comprehensive agricultural semantic resource is the multilingual AGROVOC controlled vocabulary, developed by the United Nations' Food and Agriculture Organization (FAO). Other than agricultural terms and concepts, it also contains diverse agricultural information i.e., environment, fisheries, forestry, food, etc., and supports the Linked Open Data Schema (LODS) to align it with other resources. Major semantic agricultural resources with supported languages are shown in Table 5.3 (Drury et al. 2019).

Farm Table

FarmID ▾	FarmName ▾	Location ▾	CropID ▾
1	Abbotsford	Canterbury	101
2	Bayfields	Takapau	102
3	Brookwood	Otago	103

Crops Table

CropID ▾	CropName ▾	PlantingDate ▾	HarvestingDate ▾	CropYield ▾
101	Wheat	2023-04-01	2023-07-01	450 Tons
102	Maize	2034-05-01	2023-09-01	500 Tons
103	Soyabean	2023-06-01	2023-10-01	700 Tons

Farmer Table

FarmerID ▾	FarmerName ▾	FarmID ▾
1	Robert Brown	3
2	Emily Johnson	2
3	Mary Smith	1

Equipment Table

EquipmentID ▾	EquipmentName ▾	FarmID ▾
1001	Seeder	1
1002	Tractor	2
1003	Harvester	3

Tables' Relationships

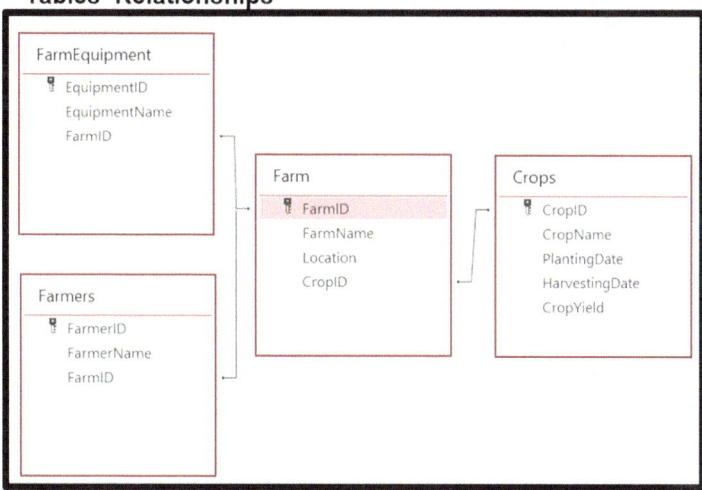

Fig. 5.2 Example of agricultural structured data (tables and tables' relationships) stored in relational database

Table 5.3 Few agricultural semantic resources and supported language(s)

Semantic resource	Supported languages
AGROVOC	Arabic, Chinese, Czech, English, French, German, Hindi, Hungarian, Italian, Japanese, Korean, Lao, Malay, Persian, Polish, Portuguese, Russian, Slovak, Spanish, Telugu, Thai, Turkish and Ukrainian
Chinese agricultural thesaurus	English and Chinese
Crop ontology	English
Cab thesaurus	English

5.4.2   Semi-structured Data

Semi-structured data refers to information that is not well-organized as it lacks a formal schema. However, it possesses some degree of structure to organize data into predefined formats with fixed data types. Key characteristics of semi-structured data include

- flexibility in data representation that makes it suitable to capture data from diverse sources
- not enforcing strict data type for its elements
- allow the use of semantic tags describing the structure of the data
- elements may be organized in a hierarchical or nested structure
- allow the combining of textual content with markup tags
- accommodates schema evolution over time to allow changes in data structure
- supports the use of query languages (JSONPath, XQuery, XPath) to extract and manipulate data.

Below are a few examples of agricultural data that can be stored in different semi-structured data formats.

- *Farm/Field Data*: Individual farm field attributes i.e., field ID, location coordinates, soil type, crop rotation history with irrigation, fertilizer, and pesticide schedules as nested properties are preferred to be stored in JSON format.
- *Weather Data*: Weather data including location name, date, temperature, humidity, atmospheric pressure, wind speed, etc. along with daily and hourly forecast details as nested properties can be stored in hierarchal JSON structures.
- *Crop Monitoring Data*: Data collected from field sensors monitoring crop growth parameters i.e., plant height, photosynthesis rates, chlorophyll content, etc. along with the location name and timestamp can be stored in JSON format.
- *Market Price Data*: JSON objects can hold market price data for agricultural commodities including commodity name, date, price, volume, etc.

```json
{
  "field_id": "C001",
  "size": 50,
  "location": {
    "latitude": 30.6319,
    "longitude": -77.1151
  },
  "soil_type": "Loam",
  "crop_rotation": ["Wheat", "Maize", "Soybeans"],
  "soil_analysis": {
    "pH": 6.6,
    "nitrogen": 45,
    "phosphorus": 25,
    "potassium": 75
  },
  "irrigation_schedule": {
    "start_date": "2024-03-15",
    "end_date": "2024-07-31",
    "frequency": "Weekly"
  },
  "pest_management": {
    "pests": ["Corn borers", "Aphids"],
    "control_methods": ["Pesticide", "Crop rotation"]
  }
}
```

```json
{
  "forecast_date": "2024-03-31",
  "location": {
    "latitude": 40.7128,
    "longitude": -74.0060
  },
  "forecast": [
    {
      "time": "01:00 PM",
      "temperature": 25,
      "precipitation": 0,
      "humidity": 63,
      "wind_speed": 15,
      "pressure": 1145
    },
    {
      "time": "3:00 PM",
      "temperature": 22,
      "precipitation": 0,
      "humidity": 55,
      "wind_speed": 12,
      "pressure": 1010
    },
  ]
}
```

Fig. 5.3 **a** Farm field data in JSON format. **b** Weather forecast data in JSON format

The actual storage of above mentioned agricultural data in different semi-structured data formats has been discussed below.

JSON (JavaScript Object Notation): JSON is a text-based semi-structured human and machine-readable data format that is used to store and transmit JavaScript objects (consisting of literals, arrays, scalar data, etc.). JSON data is stored in a hierarchical structure that is suitable to represent nested and complex datasets. Farm field data and weather forecast data in JSON format have been shown in Fig. 5.3a, b.

CSV (Comma-Separated Values): CSV files are special text files that contain tabular data where attribute values are separated by commas and records are separated by newlines. The main advantages of CSV file is its simplicity, platform independence, compatibility with other software applications like database management systems (like MySQL) and spreadsheets (like Microsoft Excel), and scalability as these files can handle millions of records without performance degradation. These attributes of the CSV format make it a preferred choice by industries like healthcare, e-commerce, finance, and agriculture. Crop Planting Schedules data and Soil Analysis Results data in CSV format have been shown in Figs. 5.4 and 5.5.

XML (eXtensible Markup Language): XML is a hardware/software independent markup language that is used to store information (in the form of hierarchical elements) that can be easily transported over the web. The main strength of XML is its support for information exchange over the web (across different computer systems, databases, and third-party applications). Farm Field data XML format has been shown in Fig. 5.6.

```
Crop Type,Planting Date,Variety,Seeding Rate (kg/ha),Planting Method
Corn,2022-04-01,Hybrid A,120,Direct Seeding
Soybeans,2022-05-15,Variety B,80,Drilling
Wheat,2022-09-01,Variety C,100,Broadcast Seeding
```

Fig. 5.4 Crop planting schedules in CSV format

```
Location,Soil pH,Nitrogen (ppm),Phosphorus (ppm),Potassium (ppm),Organic Matter (%),Texture
Field A,6.5,40,20,80,3.5,Loam
Field B,7.0,35,25,75,4.0,Sandy Loam
```

Fig. 5.5 Soil analysis results in CSV format

```xml
<agricultural_data>
    <farm>
        <farm_name>Golden Harvest Farm</farm_name>
        <location>
            <latitude>34.0522</latitude>
            <longitude>-118.2437</longitude>
        </location>
        <crops>
            <crop>
                <crop_name>Wheat</crop_name>
                <planting_date>2024-02-01</planting_date>
                <harvest_date>2024-06-15</harvest_date>
                <quantity>10000</quantity>
            </crop>
            <crop>
                <crop_name>Maize</crop_name>
                <planting_date>2024-04-15</planting_date>
                <harvest_date>2024-09-30</harvest_date>
                <quantity>8000</quantity>
            </crop>
        </crops>
    </farm>
</agricultural_data>
```

Fig. 5.6 Farm field data in XML format

To store these types of semi-structured data formats, NoSQL (Not Only SQL) databases are preferred. NoSQL databases are non-tabular and well-suitable for specific types of applications i.e., web, social media, real-time systems, etc. that have diverse storage requirements for semi-structured or unstructured data types. The main characteristics of these databases include schema flexibility, horizontal scaling, and fast querying due to the support of a variety of data models. Depending on the type of data models, different types of NoSQL databases have been developed. The most used NoSQL databases are MongoDB, Oracle NoSQL Database, Apache Ignite, etc.

5.4.3 Unstructured Data

Unstructured data refers to the information that is stored as raw files i.e., images (JPEG, GIF, PNG, etc.) videos (MP4, MPEG, AVI, etc.), audio (MP3, WAV, AAC, etc.), documents (TXT, DOC, PDF, etc.), etc. Therefore, this type of data has no defined structure. In recent times, unstructured data has played a vital role in various industries including medicine, marketing, finance, agriculture, etc. in a way that effectively managed unstructured data provides opportunities for innovation and growth. Examples of unstructured data in smart agriculture include the data collected by sensors in the form of readings (temperature, moisture, rainfall, etc.), satellite/ drone imagery, field notes and observations, weather reports, social media platforms, etc.

By leveraging structured, semi-structured, and unstructured data in digital agriculture, agriculturists can harness the power of IoT, artificial intelligence, machine learning, neural networks, etc. to improve outcomes across the entire agricultural value chain. The summary of the characteristics of all three types of data along with relevant examples of agricultural data are discussed in Table 5.4.

There are multiple options available for organizations to store BigData i.e., On-premises BigData storage (using own infrastructure and storage system), Cloud-based BigData storage (using storage space provided by cloud providers), Hybrid Storage (combination of both On-premises and Cloud storage) Approach. The choice of BigData storage depends on multiple factors, i.e., data volume, budget constraints, performance requirements, security constraints, and organizational preferences. However, in recent times, most organizations have preferred the cloud. Two main reasons for this choice are cloud storage's cost-effectiveness and its support for BigData analytics. The existence of this symbiotic relationship between BigData and the cloud complements and enhances the capabilities of each other. The next sections describe the concept of BigData storage and processing in the cloud with reference to their implications in digital agriculture systems.

5.5 The Cloud

The term Cloud refers to the infrastructure consisting of a collection of distributed servers that hosts hardware, software, and on-demand services i.e., storage and processing. As BigData is often stored in the cloud, efficient IoT-based agricultural systems often leverage cloud-based solutions (Cloud Storage and Cloud Computing)

Table 5.4 Characteristics of structured, semi-structured, and unstructured data

Characteristics	Structured data	Unstructured data	Semi-structured data
Organization and format	Organized data or fixed-format data	Unorganized data stored in an unknown format	Data not in the tabular form but in the form of tags to induce semantic elements and hierarchies of data records
Technology type	Relational databases (in rows and columns)	Not stored in relational database	Based on semantic Tags or metadata i.e., XML, HTML, JSON
Data type	Usually characters	Both characters and binary	Both characters and binary
Data organization and management	Easy to organize and manage	Difficult to organize and manage	Easier to organize and manage
Storage requirements	Requires less storage space	Requires more storage space	Required less space than unstructured data
Flexibility	Schema dependent and less flexible	Absence of schema and more flexible	More flexible than structured data but less flexible than unstructured data
Scalability	Difficult to scale	Scalability is simpler	More scalable structured data
Analysis	Easy to analyze	Difficult to analyze	Easier to analyze than structured data
Example agricultural data	Crop data, soil data, weather data, livestock data, pesticide/fertilizer usage data, market and pricing data, farm management data, etc.	Aerial imagery, sensor readings, audio/video data, images of crop diseases and pests, published articles on agriculture research, etc.	Data from agricultural websites, NoSQL databases stored in JavaScript Object Notation (JSON) or Extensible Markup Language (XML), Log files, CSV (Comma-Separated Values) Files, etc

that provide scalable and secure repositories to agriculturists for huge storage and efficient processing.

5.5.1 Cloud Storage

Cloud Storage is a model of data storage that supports the remote storage of digital data in logical pools that are established over multiple connected physical servers and accessible over the Internet. Unlike local storage, cloud storage offers greater scalability (provisioning of storage capacity on demand), accessibility (ubiquitous availability over private/public internet connections), and reliability (ability to perform and deliver services consistently). Other advantages offered by cloud storage services for BigData applications include efficient data processing and analytics. Moreover, the cloud providers are also responsible for physical server security and the protection of data from unauthorized usage. Amazon S3, Amazon Elastic File System, Google Cloud, IBM Cloud, and Microsoft Azure are popular cloud storage providers.

5.5.1.1 Cloud Storage Architecture

Cloud storage is based on virtualized infrastructure established over multiple inter-connected physical storage servers located in the same or different geographical locations. Key components of typical cloud storage architecture that work together for the provisioning of scalable and reliable storage services have been discussed below.

- *Clients*: Clients comprising end-user devices (smartphones, laptops, desktop computers) and (desktop, web, mobile) applications that interact with the cloud storage system to access data.
- *Frontend Interfaces*: these interfaces comprising of Application Programming Interfaces (APIs) enable clients to create folders, upload, download, and manage data.
- *Backend Interfaces*: these involve underlying hardware resources (servers, storage drives, network devices, etc.) and software components for data storage and management.
- *Storage Nodes*: Individual servers and storage devices are known as *storage nodes* that are organized into clusters. Data stored on storage nodes is typically distributed across multiple locations for disaster recovery purposes.
- *Metadata Service*: Cloud metadata refers to the information about the data in the cloud and it is crucial to be stored in the cloud as it enables efficient data management along with helping users to identify and categorize files. This is the metadata service at cloud that stores metadata (i.e., file names, file sizes, timestamps, and access permissions) associated with users and assists cloud service providers in organizing, maintaining, and delivering services to clients.
- *Replica Manager*: Storage clouds often employ a module known as Replica Manager that ensures the availability of cloud data. The Replica Manager period-ically analyzes data access statistics and assures data replication across multiple (appropriate) storage nodes located at different sites to protect data against service disruption and hardware failures.
- *Access Control*: To comply with privacy and regulatory requirements, access control mechanisms are required to be implemented that include authentication and authorization. Authorization by verifying the user/service identity and authen-tication by determining users' access rights ultimately protect the system and stored information, respectively.

5.5.1.2 Types of Cloud Storage

At the basic level, all cloud storage systems include above mentioned key components. However, depending on the specific requirements regarding data management, cloud storage can be centralized and distributed.

Centralized Cloud Storage: Centralized cloud storage enables data storage at a single central location of cloud infrastructure that typically comprises of one or more (physical) storage servers managed by a cloud service provider. Centralized storage is mostly cost-effective as it requires limited resources and infrastructure. Moreover, due to the central repository, administration, backup management, access control, and enforcement of security policies are easier to implement. Single point of failure, performance bottlenecks due to multiple concurrent access, and scalability challenges (i.e., significant investment and downtime) to increase capacity and power are the main disadvantages of centralized cloud storage systems.

Distributed Cloud Storage: In distributed cloud storage, data is distributed across multiple geographically dispersed data centers. Fault tolerance due to replicated data, inherent support for scalability, and improved performance by reducing the risks of bottlenecks are the main advantages of distributed cloud storage. However, design and implementation complexity (compared to centralized cloud storage systems), high-level demands of replication, and concurrency to ensure data consistency are the main disadvantages of these types of cloud storage systems.

Ultimately, the choice of centralized and distributed cloud storage system depends on various factors including size and complexity of data, storage, fault tolerance needs, etc.

5.5.1.3 Cloud Storage in Digital Agriculture

In digital agriculture, both centralized and distributed cloud storage play an essential role by providing a scalable, reliable, and accessible solution to store and manage huge amounts of agricultural data generated by deployed sensors and IoT devices in outdoor/indoor agricultural ecosystems. Below are a few advantages of cloud storage implications in digital agriculture.

Ubiquitous Accessibility: The seamless anywhere accessibility to agricultural data over the internet is one of the main advantages of cloud storage. It enables agricultural stakeholders to monitor and analyze regardless of their current location.

Scalability: The offering of virtually unlimited storage capacity allows agricultural organizations, companies, and institutions to accommodate the increasing volumes of agricultural data generated by sensors and IoT devices without significant upfront investments in infrastructure.

Data Redundancy and Disaster Recovery: The replication of data by different agricultural organizations across multiple geographic regions minimizes the risk of data loss by cyber-attacks, hardware failures, natural disasters, etc.

Collaboration and Sharing: By collaborative workspaces of cloud storage platforms, agriculturists (i.e., farmers, researchers, agronomists, entomologists, plant pathologists, pesticide and fertilizer companies, and policymakers in government can securely share data and insights with each other.

Cost-effectiveness: Agricultural organizations can save upfront investments and operational costs with the pay-as-you-go pricing model of cloud storage.

Combining the potential of centralized and distributed storage clouds, stakeholders of digital agriculture can harness the power of cloud computing to increase crop productivity and promote sustainable farming.

5.5.2 Cloud Computing

In general, cloud computing can be considered a colossal data center with the provisioning of online services at both public and private levels (Ayaz et al. 2019). The three common layers of cloud computing architecture are Infrastructure-as-a-Service (IaaS), Platform-as-a-Service (IaaS), and Software-as-a-Service (IaaS). The layered cloud service model in digital agriculture is shown in Fig. 5.7 (Nath and Chaudhuri 2012). From Fig. 5.7, it becomes evident that the layered cloud service model of digital agriculture encompasses different functionalities to cater diverse needs of agricultural stakeholders.

5.5.2.1 Infrastructure-as-a-Service (IaaS)

IaaS supports the on-demand availability and management of physical resources through virtualization techniques. In other words, this layer provides virtualized

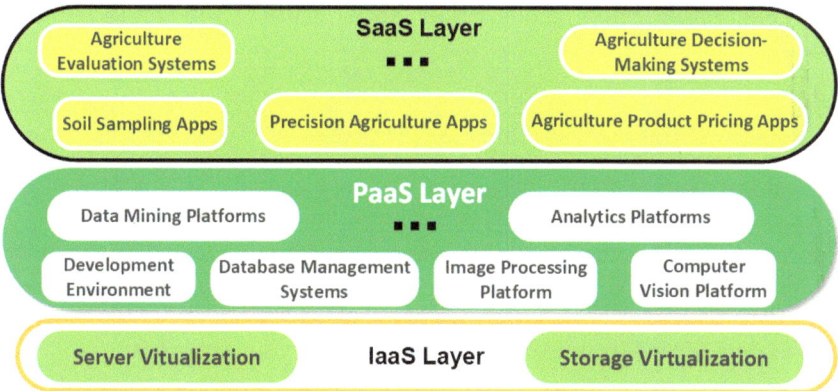

Fig. 5.7 Cloud layered service model in digital agriculture

storage and computing resources on a pay-as-you-go basis. In digital agriculture, IaaS offerings assist agricultural organizations and institutions to leverage computing infrastructure without the need for upfront investment to buy physical (computing) resources and management of complex computing environments. This computing infrastructure enables agriculturists to host agricultural applications, store data, and perform analytics.

5.5.2.2 Platform-as-a-Service (PaaS)

PaaS supports the availability and execution of remote instances of system software (operating system) and development software packages. The development software packages assist in the development, testing, and deployment of applications. In digital agriculture, PaaS layer plays a vital role with the provisioning of development and deployment platforms to build, test, and deploy agricultural tools and applications. These tools and applications include data mining platforms, analytic platforms, database management systems, computer vision, image processing platforms, etc. that enable innovations in crop management, livestock management, and supply-chain optimizations.

5.5.2.3 Software-as-a-Service (SaaS)

SaaS supports the delivery of ready-to-use instances of software applications, systems, and services over the Internet on a pay-as-you-go basis. Within the context of digital agriculture, these applications and services include the support of various agriculture applications (i.e., crop monitoring, product pricing, weather forecast, livestock tracking, etc.) and systems (i.e., land evaluation systems, agriculture expert systems, agriculture decision-making systems, etc.).

In summary, concerning the requirements of digital agriculture, cloud technology offers many opportunities to farmers, food companies, pesticide and fertilizer organizations including

- aggregation of field data i.e., data from soil sensors, satellite images of agricultural fields, fertilizer or pesticide usage information in a particular area.
- better decision-making through the usage of agricultural knowledge-based massive-scale repositories that contain information about traditional and innovative farming practices and experiences.
- support data analytics to improve agriculture production.
- support information sharing that helps researcher communities (scientists, students, faculty) to share discoveries and suggestions about the implementation of modern techniques of crop cultivation and livestock management.

Despite all its advantages, the cloud storage/computing paradigm is not suitable for all types of applications especially for delay-sensitive applications because resources in cloud data centers are centralized and generally located far away from end-user devices. For example, in the case of latency-critical applications, it is impractical to send data to the faraway cloud for storage and processing. Excessive (network and cloud) resource utilization while uploading all sensed data to the cloud (that rarely changed), lack of real-time data evaluation, excessive dependency on the network, and high risks to data security and privacy are the main disadvantages of cloud computing. These problems also increase the cost of digital agriculture systems in terms of high bandwidth, storage, and computing requirements (Zhang et al. 2020). Edge computing and Fog computing technologies have resolved these issues through the provisioning of data processing closer to the initial data generated site i.e., at the network edge and network core (Garcia Lopez et al. 2015; Shi et al. 2016).

5.6 Edge Computing and Fog Computing

Edge Computing supports the processing or computation of data where or near where the data is produced. In other words, Edge computing provides local processing at the edge or near the network edge. This local processing near the end-users mitigates the computational stress of the cloud and reduces the latency of response time in digital systems (Salman et al. 2015; Varghese et al. 2016). End devices (i.e., smart things, smart/mobile phones, etc.) and edge devices (i.e., switches, bridges, routers, base stations, wireless access points, etc.) equipped with specialized capabilities support edge computation. Due to this localized processing capability, Edge computing provides a faster response to service requests and mostly resists sending raw data to the core network.

Fog computing besides enabling edge computation at the network edge can be expanded to the network core as well (Bonomi et al. 2012). In simple words, Fog computing's computational infrastructure can be categorized as fog devices, fog servers, and gateways. Fog devices (also known as the edge network component of fog computing) deployed closer to the data source including IoT devices, routers, switches, etc. Fog servers are powerful computing nodes (typically located closer to the edge devices than cloud servers) that act as intermediaries between edge devices and cloud servers. These nodes provide additional storage capacity and computing power to support complex data analytics tasks. Fog gateways are the devices that are responsible for redirecting information between fog devices and servers. Like Edge computing, the edge network component of Fog computing placed closer to sensors, smart things, and IoT devices enables data storage and processing within the local vicinity to improve service delivery latency for real-time applications. However, unlike Edge computing, Fog servers provide cloud-based services i.e., IaaS, PaaS, and SaaS. Due to this Fog computing has been regarded as a more well-structured paradigm as compared to Edge computing. A brief comparison of Edge, Fog, and Cloud technologies has been shown in Table 5.5.

Table 5.5 Comparison between fog and cloud characteristics

Cloud/fog characteristics	Edge computing	Fog computing	Cloud computing
Scalability	Limited	Limited	Dynamic and high
Latency	Low	Medium	High
Location-awareness	Yes	Yes	No
Real-time response	Yes	Yes	No
Storage capacity	Low	Medium	High

5.6.1   Agricultural Use Cases of Edge/Fog Computing

Self-driving tractors and automatic (or robotic) machinery are used in agriculture farms to communicate with deployed field sensors. Using computer vision techniques and pre-loaded field data at the edge, these machines able to

- calculate the most effective paths to cover the required piece of land
- find new paths in case of an unexpected obstacle
- perform watering, weeding, and harvesting of crops.

Complete automation of closed ecosystems i.e., greenhouses fundamentally relies on an edge computing approach to managing various processes i.e., temperature control, light control, plant watering, cattle feeding, etc.

Prediction of potential environmental hazards and disasters through the use of localized (edge) resources in the digital agriculture systems enables farmers to take immediate measures to protect the crop, plants, and or livestock.

In all the above-mentioned use cases of digital agriculture systems, edge computing assists the automation of required operations regardless of the connectivity of farm/greenhouse devices to the main server and makes decisions locally. In addition to the implications of edge/fog computing use-cases in digital agricultural systems, the following are specific agriculture use cases where the coordination of Edge/Fog computing with cloud computing technology helps to meet the global challenges of agricultural sustainability and food shortage (Zamora-Izquierdo et al. 2019; O'Grady et al. 2019; Jukan et al. 2019).

- Farm automation
- Monitoring of farm environments
- Health and welfare of livestock
- Crop production and forestry
- Aquafarming, vertical farming
- The food chain supply.

5.6.2 Example of Fog Computing Advantage Over Cloud Computing

The significance of Edge/Fog computing over Cloud computing has been explained below with the help of a typical scenario depicting the integrated use of Edge/Fog computing with Cloud computing as shown in Fig. 5.8. Consider one of the setups shown in Fig. 5.8, where an agricultural robot (named "Identify & Treat" robot) has taken an image of a plant leaf intending to identify and treat the plant disease. Suppose the size of the image is 32 MB (Mega-Bytes) which needs to be processed within one second on remote servers. Assume the same amount (32 MB) of processed data is sent back to the robot after processing. Which option would you choose to meet this timing constraint? Explain your reasoning.

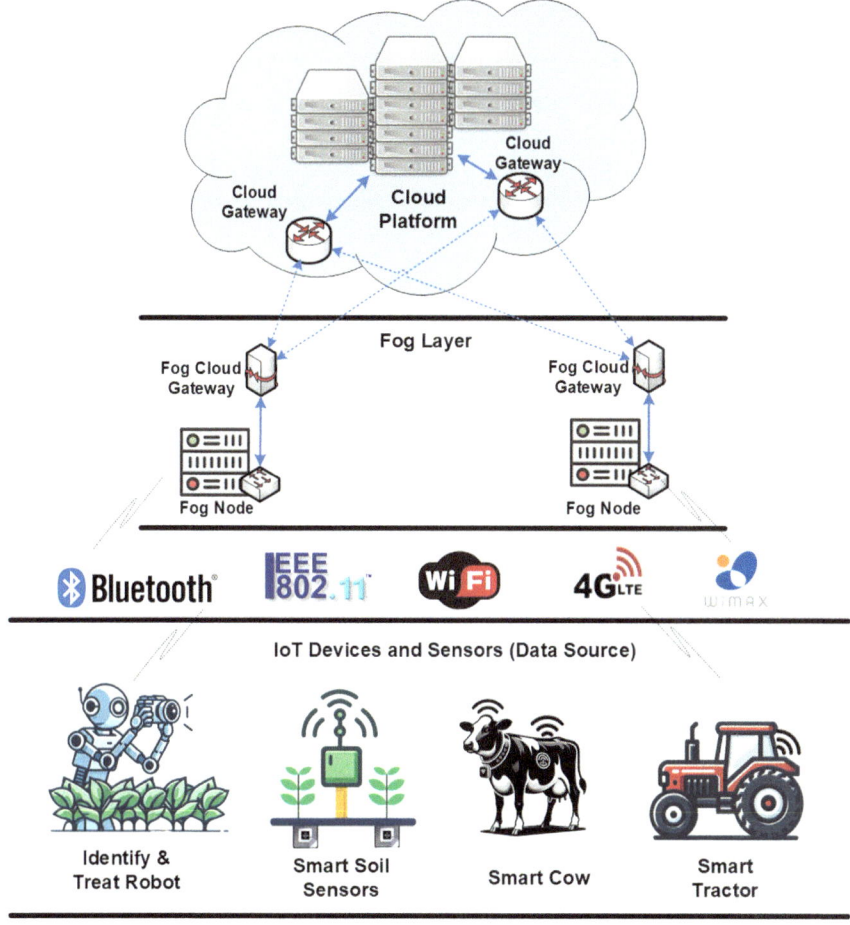

Fig. 5.8 Agricultural IoT with edge/fog and cloud computing implementations

(a)   Sending the image to a Fog server (for disease identification) over a duplex link
      with a data rate of 128 MB/s (Mega-Bytes/second) and a Propagation Delay of
      25 ms. The processing of the data on the Fog server takes 10 ms.
(b)   Sending the image to a Cloud server (for disease identification) over a duplex
      link of 128 MB/s and a Propagation Delay of 300 ms. The processing of the
      data on the Cloud server takes 10 ms.

Solution

Size of the image in bits $(L_1) = 32$ MB $= 32 \times 8 \times 10^6$ bits.
   Size of response data in bits $(L_2) = 32$ MB $= 32 \times 8 \times 10^6$ bits.
   Duplex link data rate for Fog server $(R_f) = 128$ MB/s
   Duplex link data rate for Cloud server $(R_c) = 128$ MB/s
   The generic formula to calculate the overall delay from source to destination and
back from destination to the source is given in Eq. 5.1.

$$
\begin{aligned}
\text{Overall Delay} \; = \; & \text{Transmission Delay1} \; (T_{d1}) + \text{Propagation Delay1} \; (P_{d1}) \\
& + \text{Processing Delay} \; (P_{rd}) + \text{Transmission Delay 2} \; (T_{d2}) \\
& + \text{Propagation Delay 2} \; (P_{d2}) \quad\quad\quad\quad\quad\quad\quad (5.1)
\end{aligned}
$$

The formula to calculate Transmission Delay $(T_{d1}$ or $T_{d2})$ is given in Eq. (5.2).

$$
\text{Transmission Delay} \; (T_d) \; = \text{Data Size in bits} \; (L)/\text{Data rate on the link (bits/s)} \tag{5.2}
$$

T_{d1} and T_{d2} in Eq. 5.1 represent transmission delay from sender (robot) to receiver
(Fog or Cloud server) and from destination (Fog or Cloud server) to source (robot),
respectively.
   Processing Delay (P_{rd}) is fixed but varies from device to device. Below, we have
mentioned it as P_{rdf} (to represent the processing delay on the Fog server) or P_{rdc} (to
represent the processing delay on the Cloud server). In this example, both Fog and
Cloud servers have the same processing delay which is 10 ms.
   The formula to calculate Propagation Delay $(P_{d1}$ or $P_{d2})$ is given in Eq. (5.3).

Propagation Delay (P_d)
$=$ Link distance from sender to the receiver in meters $(D)/$Speed of link (m/s) (5.3)

P_{d1} and P_{d2} in Eq. 5.1 represent propagation delay from sender (robot) to receiver
(Fog or Cloud server) and from receiver (Fog or Cloud server) to sender (robot),
respectively. In this example, these delays have already been given in the problem
statement.

(a)   *Overall delay in the case of Fog computing*

Transmission Delay in the case of sending data from the robot to the Fog server is

$T_{d1} = L_1/R_f = (32 \times 8 \times 10^6 \text{ bits})/(128 \times 8 \times 10^6 \text{ bits/s}) = 0.25$ s.

Propagation Delay from the robot to the Fog server (P_{d1}) = 25 ms = 0.025 s.

This P_{d1} (0.025 s) is given in the problem statement (that was actually calculated by considering the formula given in Eq. 5.2).

Processing Delay on Fog Server (P_{rdf}) = 10 ms = 0.01 s.

Transmission Delay in the case of sending data from the Fog server to the robot is

$T_{d2} = L_2/R_f = (32 \times 8 \times 10^6 \text{ bits})/(128 \times 8 \times 10^6 \text{ bits/s}) = 0.25$ s.

Propagation Delay from the Fog server to the robot (P_{d2}) = 25 ms = 0.025 s.

This P_{d2} (0.025 s) is given in the problem statement (that was actually calculated by considering the formula given in Eq. 5.2).

Overall Delay = $T_{d1} + P_{d1} + P_{rdf} + T_{d2} + P_{d2} = 0.25 + 0.025 + 0.01 + 0.25 + 0.025 = 0.56$ s.

(b) *Overall delay in the case of Cloud computing*

Transmission Delay in the case of sending data from the robot to the Cloud server is

$T_{d1} = L_1/R_c = (32 \times 8 \times 10^6 \text{ bits})/(128 \times 8 \times 10^6 \text{ bits/s}) = 0.25$ s.

Propagation Delay from the robot to the Cloud server (P_{d1}) = 300 ms = 0.3 s.

This P_{d1} (0.3 s) is given in the problem statement (that was actually calculated by considering the formula given in Eq. 5.2).

Processing Delay on Cloud Server (P_{rdc}) = 10 ms = 0.01 s.

Transmission Delay in the case of sending data from the Cloud server to the robot is

$T_{d2} = L_2/R_c = (32 \times 8 \times 10^6 \text{ bits})/(128 \times 8 \times 10^6 \text{ bits/s}) = 0.25$ s.

Propagation Delay from the Cloud server to the robot (P_{d2}) = 300 ms = 0.3 s.

This P_{d2} (0.3 s) is given in the problem statement (that was actually calculated by considering the formula given in Eq. 5.2).

Overall Delay = $T_{d1} + P_{d1} + P_{rdc} + T_{d2} + P_{d2} = 0.25 + 0.3 + 0.01 + 0.25 + 0.3 = 1.11$ s.

Fog computing implementation would be preferred in this case, because the overall delay in the case of Fog computing fulfills the requirements of the mentioned constraint of receiving a response within 1 s.

Questions

Q5.1: Non-stop monitoring of all the cows' activities on a dairy farm is required to achieve high and good quality milk production. To address this challenge, an IoT-based solution is proposed that not only helps in monitoring the cows' activities at the farm but also facilitates farmers to learn about cows' behavior through ML-based solutions implemented in the cloud. Considering the implementation perspective of this type of IoT system, is there any need to consider for an edge computing system to be deployed? If yes, with the help of an example scenario, describe how an edge-based solution can affect the efficiency of the system in comparison to a cloud computing solution.

Q5.2: Consider a smart farming scenario where a farmer is using a mobile application (based on cloud computing technology) that has been developed for the identification of crop pests in different areas of the agriculture field. Draw the architecture diagram of the developed mobile application system for recognizing pests using some machine learning techniques implemented in a cloud computing system.

Q5.3: Provide two scenarios of precision farming where fog computing plays a critical role in data processing and decision-making.

Q5.4: How does edge computing enhance real-time decision-making and efficiency in smart farming?

Q5.5: Consider an agriculture scenario, where one of the smart farming applications running on a driverless tractor has generated 256 MB of image data, which needs to be processed as quickly as possible at remote servers. Assume the processed data (of size 32 MB) is sent back to the application after processing, Which option would you choose to meet this timing constraint? Explain your reasoning.

a. Sending the data to a Fog node over a duplex link with 128 MB/s and 200 ms propagation latency. The processing of the data on this Fog device takes 10 ms.
b. Send the data directly to the Cloud over a duplex link of 128 MB/s and 300 ms propagation latency. The processing of the data on the Cloud takes 5 ms.

References

Antoniou G, Van Harmelen F (2004) Web ontology language: Owl. Handbook on ontologies. Springer, pp 67–92

Ayaz M et al (2019) Internet-of-Things (IoT)-based smart agriculture: toward making the fields talk. IEEE Access 7:129551–129583

Bonomi F et al (2012) Fog computing and its role in the internet of things. In: Proceedings of the first edition of the MCC workshop on Mobile cloud computing

Carbonell I (2016) The ethics of big data in big agriculture. Internet Policy Rev 5(1)

Cravero A et al (2022) Data type and data sources for agricultural big data and machine learning. Sustainability 14(23):16131

Drury B et al (2019) A survey of semantic web technology for agriculture. Inf Process Agric 6(4):487–501

Garcia Lopez P et al (2015) Edge-centric computing: vision and challenges. ACM New York, NY, USA

Jukan A et al (2019) Fog-to-cloud computing for farming: low-cost technologies, data exchange, and animal welfare. Computer 52(10):41–51

McGuinness DL, Van Harmelen F (2004) OWL web ontology language overview. W3C Recommendation 10(10):2004

Miller E (1998) An introduction to the resource description framework. Bull Am Soc Inf Sci Technol 25(1):15–19

Nath B, Chaudhuri S (2012) Application of cloud computing in agricultural sectors for economic development. In: Interplay of economics, politics and society for inclusive growth-international conference organized by RTC and GNHC

O'Grady M, Langton D, O'Hare G (2019) Edge computing: a tractable model for smart agriculture? Artif Intell Agric 3:42–51

Pan JZ (2009) Resource description framework. Handbook on ontologies. Springer, pp 71–90

Posadas BB, Gilbert JE (2020) Regulating big data in agriculture. IEEE Technol Soc Mag 39(3):86–92

Sagiroglu S, Sinanc D (2013) Big data: a review. In: 2013 international conference on collaboration technologies and systems (CTS)

Salman O et al (2015) Edge computing enabling the Internet of Things. In: 2015 IEEE 2nd world forum on internet of things (WF-IoT)

Shi W et al (2016) Edge computing: vision and challenges. IEEE Internet Things J 3(5):637–646

Varghese B et al (2016) Challenges and opportunities in edge computing. In: 2016 IEEE international conference on smart cloud (SmartCloud)

Wolfert S et al (2017) Big data in smart farming–a review. Agric Syst 153:69–80

Zamora-Izquierdo MA et al (2019) Smart farming IoT platform based on edge and cloud computing. Biosys Eng 177:4–17

Zhang X, Cao Z, Dong W (2020) Overview of edge computing in the agricultural internet of things: key technologies, applications, challenges. IEEE Access 8:141748–141761

Chapter 6
Data Analytics in Digital Agriculture

6.1 Learning Objectives

After studying this chapter, students will be able to

- describe the basic enabling techniques and technologies of data analytics and data visualization
- elaborate different categories of BigData analytics along with their implications in the field of agriculture
- demonstrate the fundamental statistical and machine learning-based analysis techniques and technologies along with their applications in agriculture.

6.2 Data Analytics and Digital Agriculture

In recent times, Data is one of the most vital assets of any organization as its efficient use assists businesses to increase revenue with improved effective decision-making. The efficient use of data to discover meaningful insight for accurate predictions ultimately demands Data Science and Data Analytics services. Data Science is a broad-term and multidisciplinary field that deals with collecting, processing, analyzing, and interpreting complex datasets through exploratory data analysis using statistical/mathematical modeling, machine/deep learning approaches, and data visualization with primary goals of knowledge extraction from data and generation of actionable insights (Dhar 2013). Data analytics is the subset of data science and is specifically focused on analyzing data to discover insights. The main difference between data science and data analytics is the scope of their application. Data science encompasses the entire process of data collection, processing, analysis, and interpretation but data analytics fundamentally deals with the analysis and interpretation phases of data lifecycle. While data analytics involves datasets of varying sizes (small to medium), BigData analytics at the same time deals with the analysis of large and complex datasets. However, similar to data analytics while handling the complexity and scale of data, BigData analytics also involves a wide range of advanced data analysis methods and techniques that can be categorized as descriptive analytics (understanding of what happened), diagnostic analytics (elaborating why it happened), predictive analytics (forecasting of what will happen), and prescriptive analytics (suggesting actions to take) (Vassakis et al. 2018). Various techniques of BigData analytics under these categories have been shown in Fig. 6.1 along with the relationship between BigData and data science.

Figure 6.1 also depicts that BigData analytics emerges when data science techniques are employed on BigData. Therefore, BigData analytics can be defined as the process of discovering patterns, trends, correlations, relationships, and insights in huge volumes of data that are not possible with traditional data analytics approaches and tools. Traditional data analytics takes place after a certain time period (e.g., daily, weekly, monthly, or yearly sale analysis of a shop) but BigData analytics usually take place in real-time (e.g., route optimization by using huge amounts of data regarding road traffic, vehicles' speed, vehicles' locations, etc.). From a digital agriculture perspective, the application of BigData analytics is essential and beneficial for agriculturists to accurately predict the agricultural risks related to the production, processing, management, market, finance, supply chain management,

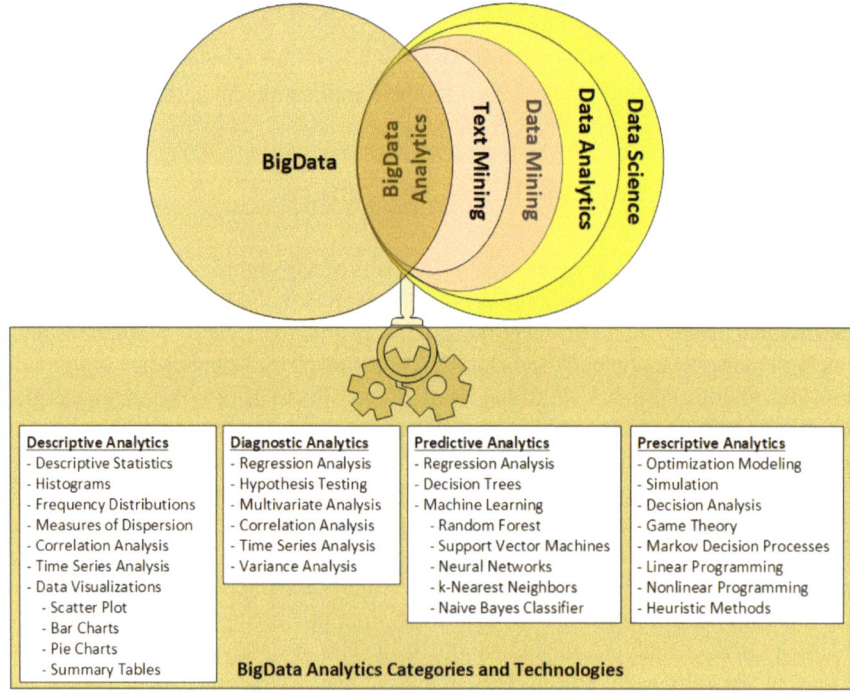

Fig. 6.1 Techniques under different categories of BigData analytics

etc. for a certain agricultural product at a particular instant of time in a specific farm or piece of farmland (Wolfert et al. 2017; Weersink et al. 2018; Liu et al. 2020). Table 6.1 describes the sources of agricultural BigData and corresponding techniques to perform BigData analytics (Kamilaris et al. 2017).

6.3   BigData Analytics Categories and Digital Agriculture

It is already mentioned in Sect. 6.2 that there are four categories of BigData analytics i.e., Descriptive Analytics, Diagnostic Analytics, Predictive Analytics, and Prescriptive Analytics. This Section describes the details regarding the implications of these data analytical techniques in the context of digital agriculture.

6.3.1   Descriptive Analytics

Descriptive analytics is the most simple and common method of data analysis that involves the understanding of "what happened" or "what are main data characteristics" by analyzing historical data. The focus of descriptive analytics revolves around the finding of patterns, trends, and correlations that can help scientists to get deeper insight. Moreover, descriptive analytics assists in presenting the summarized data in a meaningful way (e.g., revenue reports in tabular form) that is easy for concerned people to read and understand. Common approaches include summary statistics, charts and graphs, frequency and pivot tables, data visualizations, etc.

Agriculture Use Case

Descriptive analytics plays a vital role in digital agriculture through the provisioning of valuable insights into past performance, trends, and patterns that are essential to increase farm production, optimize farm operations, and improve profitability. Summarizing agricultural data including crop yield, water usage, soil moisture levels, temperature variations in indoor/outdoor farmlands, pest infestations, fertilizer/pesticide usage, etc. helps agriculturists to get a better understanding of overall farm performance. In this way, farmers can make proper adjustments that are required to improve farm productivity. On the other hand, data visualizations (in the form of charts and graphs) assist agriculturists in identifying crop yield variations, water pattern fluctuations, anomalies, and outliers in collected farm data. The identification of these anomalies at an early stage enables farmers to take proactive measures to mitigate risks and optimize farm performance. Moreover, descriptive analytics facilitates agriculturists to benchmark their farm output against industry benchmarks and ultimately enables them to set realistic goals.

Table 6.1   BigData sources and corresponding BigData analytics platforms/techniques

Agricultural entity	BigData sources	Platforms and techniques for BigData analytics
Land	• Land characterization dataset (Soil properties, topography, vegetation, land use, land cover, hydrology, climate rainfall, etc.) • Remote sensing data (multispectral satellite imagery) • Weather stations data	• Cloud platforms • Statistical analysis • Image processing • Machine learning
Weather and climate change	• Weather stations data • Remote sensing data (earth observation data from satellite) • Geospatial data [from field data collection, satellites, geographical information science (GIS)]	• Cloud platforms • Statistical analysis • Machine learning • MapReduce analytics • GIS geospatial analysis
Soil	• Ground sensors (moisture, salinity, electrical conductivity)	• Cloud platforms • Machine learning
Crops	• Ground sensors (metabolites) • Remote sensing (satellite) • Historical datasets (land use, national land information • Statistical data (on yields)	• Statistical analysis • Machine learning • Image processing (Fourier transform, Wavelet transform filtering)
Weeds	• Remote sensing data with UAVs • Normalized difference vegetation index (NDVI) sensors • Historical information (digital library of images of plants and weeds, plant-specific data)	• Machine learning (logistic regression) • Deep learning • Image processing • NDVI measurements
Livestock	• Livestock datasets (physiological characteristics) • Sensors (movement, heat, milk production, grazing activity, feed intake, sound) • Remote sensing data (satellite and drones' data about animal tracking) • Camera sensors (multispectral images and videos)	• Machine learning (decision trees, neural networks, scalable, vector machines)

6.3.2 Diagnostic Analytics

Diagnostic analytics answering "why it happened" and "which factors contribute to the outcome?" deals with the identification of root causes of anomalies and outliers in data. The diagnostic techniques i.e., hypothesis testing, correlation analysis, regression analysis, etc. help in finding hidden patterns and relationships in data to explain the ultimate reasons for past outcomes.

Agriculture Use Case

Diagnostic analytics plays an important role in digital farming through the provisioning of deeper insight into the factors that influence farm outcomes. Below are the implications of diagnostic analytics in digital agriculture.

- By analyzing data collected from sensors, IoT devices, drones, and satellites agriculturists can diagnose crop health issues. The data correlations among different factors i.e., nutrient levels, temperature, pest activity, etc. indicating problematic patterns and trends help agriculturists to identify potential issues i.e., pest infestation, disease attack, and nutrient deficiencies that affect crop production.
- By identifying correlations in factors i.e., crop management techniques, irrigation practices, weather conditions, soil quality, etc., agriculturists can determine the root causes of yield variability and take proper actions to improve farm production.
- The identification of inefficiencies regarding the utilization of farm resources (i.e., water, fertilizer, pesticide, etc.) by diagnostic analysis helps agriculturists to optimize resource allocation that ultimately improves environmental sustainability and cost.
- By analyzing patterns in data collected from sensors and maintenance logs, agriculturists can identify and predict equipment malfunctioning and in-time maintenance needs.
- Agriculturists through diagnostic analysis can identify existing correlations between weather patterns and yield outcome.

6.3.3 Predictive Analytics

Predictive analytics answering "what will happen" or "what are the probabilities of future happenings" involves the identification of future trends in collected data through different techniques i.e., time-series forecasting, regression analysis, and machine learning algorithms.

Agriculture Use Case

Predictive analysis plays a crucial role in digital agriculture by enabling agriculturists to mitigate future risks and take proactive decisions that are essential to improve sustainability and farm production. The applicability of predictive analytics in digital agriculture has been discussed below.

- Based on historical and real-time data of various factors i.e., crop management practices, soil characteristics, weather conditions, etc., predictive models allow farmers to make informed decisions about planting, irrigation, and harvesting.
- Based on historical weather and climate data, Predictive analytics enables agriculturists to take protective measures i.e., adjustment of planting, irrigation, and harvesting schedules against the forecasted likelihood of extreme weather conditions i.e., storms, floods, droughts, etc.
- By identifying correlations and patterns in historical data on pest populations and disease prevalence, predictive analytics assists farmers in implementing timely interventions (i.e., use of disease-resistant crop varieties and pest control measures) against the forecasted pest or disease outbreaks.
- Predictive analytics helps farmers to optimize resource allocation by forecasting the demands of farm resources i.e., labor, machinery usage, water, fertilizer, pesticide, etc. that ultimately maximizes profitability.

6.3.4 Prescriptive Analytics

Prescriptive analytics answering "what actions should take?" or "what is the best course of action?" involves data analysis to recommend specific actions to optimize outcomes through simulation modeling, algorithm optimization, decision trees, etc.

Agriculture Use Case

Prescriptive analytics in digital agriculture enables agriculturists to make informed decisions, adapt to changing conditions, and optimize agricultural practices with the provisioning of actionable insights and recommendations. Here's how prescriptive analytics is applied in digital agriculture:

- Through prescriptive analytics farmers can optimize the usage of farm inputs i.e., water, pesticide, fertilizer, etc. that ultimately maximize resource efficiency, minimize waste, and reduce cost while maintaining crop health and productivity.
- Prescriptive analytics provides the most effective strategies for crop management practices i.e., planting schedule, pest management, crop rotation, etc. that ultimately optimize crop yields and enhance environmental sustainability.
- It helps agriculturists to mitigate risks associated with labor availability, regulatory changes, weather variability, market fluctuations, etc. by recommending efficient risk management strategies i.e., diversification of crops or markets, crop insurance coverage, etc.

6.4 Data Mining and Digital Agriculture

One of the core subsets of data analytics is data mining which encompasses the process of extracting patterns, trends, and insights from large datasets using statistical and machine learning techniques. The extracted information and insights are helpful in driving informed decision-making across various industries including agriculture.

6.4.1 Data Mining

Data Mining (also known as Knowledge Discovery in Data) is the process of searching and analyzing huge volumes of data to discover anomalies, correlations, trends, patterns, etc. that ultimately improve customer relationships, reduce business risks, decrease production costs, (Ye 2003) etc. Therefore, focusing on large datasets, data mining is highly effective with the support of several statistical (Regression Analysis, Correlation, etc.) and Machine Learning (supervised, unsupervised, neural network) techniques (Han et al. 2022). Agricultural datasets are inherently diverse in nature and through the application of data mining techniques, agricultural organizations (both public and private) are able to get descriptive as well as predictive information that helps agriculturists in decision-making (Issad et al. 2019). Table 6.2 provides examples showing how data mining is playing a critical role in digital farming.

6.4.2 Text Mining

A special type of data mining approach is text mining (also known as Text analytics). Formally, Text Mining can be defined as the knowledge-intensive process in which a user interacts with a collection of documents over time using different analysis techniques (Feldman and Sanger 2007; Allahyari et al. 2017). Therefore, text mining approaches enhance the agricultural decision-making process by providing valuable insights from unstructured textual data available in different formats i.e., research articles, social media posts, weather reports, etc. (Drury and Roche 2019) as mentioned in Table 6.3.

Three core Text Mining techniques i.e., Information Retrieval, Information Extraction, and Sentiment Analysis, etc. have been discussed below with reference to the context of agriculture.

Information Retrieval deals with the finding and ranking of unstructured (text) documents that are required against specific information within large collections available on computing devices (Manning et al. 2010). Information retrieval requires topic-oriented document collection (corpus) for the systems to return pertinent documents. A few information retrieval techniques have been discussed below.

Table 6.2 Role of data mining in digital agriculture

Agricultural goal	Factors for analysis	Major data mining technique(s)	Benefit
Crop yield prediction	Historical data of crop type, soil, weather conditions	• Statistical correlation • Machine-learning-based Pattern Identification	Make optimum crop management and financial decisions
Irrigation management	Data on crop water requirements, soil moisture level, and weather forecast	• Supervised/ unsupervised machine learning • Deep learning	Reduce water wastage and conserve water to maintain crop health
Pest and disease detection	Sensor data and drone imagery	• Machine/Deep Learning Algorithms	Help to take timely preventive measures to minimize crop loss
Livestock management	Livestock health and behaviour	• Machine learning models	Ensure animal welfare by predicting disease outbreaks and optimizing feeding practice
Market analysis and demand forecasting	Consumer preferences, market data, and economic trends	• Machine learning techniques	Make informed decisions about crop production volumes and market timing to reduce market risks
Supply chain optimization	Logistics data, transportation routes, and inventory levels	• Statistical and machine learning models	Reduce costs and improve the efficiency of food production and distribution

- Natural Language Processing (NLP) techniques e.g., named entity recognition, and part-of-speech tagging allow information retrieval from unstructured data.
- Boolean Retrieval technique based on the use of Boolean operators (AND, OR, and NOT) to retrieve documents matched with specific criteria.
- Keyword Search approach used to retrieve documents matched with keywords or phrases provided by users.
- Vector Space Modeling technique enables applications to determine document relevancy by the use of documents and queries that are represented as vectors.
- Supervised and Unsupervised machine learning algorithms improve information retrieval relevance and accuracy by learning patterns and relationships in data.

In digital agriculture, retrieval of pertinent information from large volumes of agricultural data (available in the form of database tables, website content, scientific literature in the form of blogs, research articles, patents, etc.) helps agriculturists to make informed decisions about

- selection of crops, cultivation methods,

Table 6.3 Role of text mining in digital agriculture

Agricultural goal	Type of text data	Benefit
Market analysis	Online reviews, social media posts, market reports	Insights are useful to understand market trends, consumer preference and adjustments of marketing strategies
Regulatory compliance and policy development	Government reports, regulatory and policy documents	Agriculturists can adapt their practices accordingly to remain compliant with updated policies relevant to subsidies and incentives
Weather forecasting and risk management	Weather-related data available in textual format	Agriculturists can implement proactive measures to mitigate the impact of anticipated weather-related risks i.e., floods, droughts, extreme temperatures, etc
Pest and disease monitoring	Research articles, disease databases	Farmers can take preventive measures such as early action to do pesticide spray
Research collaboration	Research databases containing scientific literature in the form of articles and patents	Agricultural researchers can identify research gaps, share knowledge, and foster collaboration

- strategies of planting and harvesting, fertilizer application and pest management and control,
- mitigation of weather-related risks,
- livestock management practices in terms of animal health, breeding practices, nutrition guidelines, and veterinary protocols. etc.
- regulatory requirements, legal risks, compliance obligations, industry standards, etc.

Information Extraction approach is used to extract structured information from unstructured or semi-structured data sources. In other words, information extraction helps to transform raw textual or multimedia data into structured formats that are useful in analysis and decision-making. It identifies a subset of information within a document or sometimes it can be considered as the classification of words/multi-word expressions into predetermined categories (Okurowski 1993). Various approaches of information extraction i.e., Named Entity Detection, Monitoring, and Knowledge Extraction play important roles in different domains of digital agriculture (Drury and Roche 2019) i.e., farm management (Liao et al. 2015), animal disease detection (Goel et al. 2018), and food price prediction (Chakraborty et al. 2016; Kim et al. 2017).

With the Named Entity Detection approach (Kumar et al. 2016; Malarkodi et al. 2016), a number of named entity taggers have been developed for agriculture (Kumar et al. 2016), which ultimately rely on either machine learning or linguistic rules for the detection of named entities on the basis of similarity in training data or manually constructed rules.

Monitoring approaches (Arsevska 2017) of text mining are used to discover the evolution of specific agricultural phenomena published on the web e.g., monitoring of animal diseases (Goel et al. 2018). In this way, for livestock management, the monitoring approach ultimately assists government authorities in limiting the spread of animal disease outbreaks i.e., bird-flu, swine flu, blue tongue disease, etc. (De Quincey and Kostkova 2009; Arsevska et al. 2018). Most of the monitoring systems developed so far have used published news on the Internet or social media with a few exceptions that use autopsy reports to extract information (Küker et al. 2018).

Knowledge extraction is fundamentally the usage of information available in the text to create representations of knowledge that can be transformed into models (graph representations) to infer new knowledge. Concerning the agriculture domain, these graph representations can be used to identify relationships between entities and concepts i.e., crop and soil or crop and disease. One of the examples is the development and implementation of the Bayesian Network for Sugarcane to extract causal relationships between events (Drury et al. 2016). Regarding farm management systems, text mining assists in the decision-making process through the information extraction from literature text in the form of scientific reports related to planting, pesticide spray, fertilizer application, crop harvesting, etc. (Liao et al. 2016). Moreover, the prediction of the daily prices of food products (i.e., beef, vegetables, chicken, onion, etc.) is possible through the Nowcasting approach (or short-term price prediction) by monitoring tweets (Kim et al. 2017).

Sentiment Analysis (or Opinion Mining) has been defined as the computational study to draw meaning from text by analyzing people's opinions, attitudes, appraisals, and emotions available in the form of online reviews, customer feedback, and social media conversation towards individuals, issues, products, topics, events, etc. (Liu and Zhang 2012). There are several computing techniques that can identify and understand the relationships of entities in text documents by analyzing a language's grammatical structure i.e., text classification with logistic regression, supervised machine learning, and neural networks. In digital agriculture, sentiment analysis is helping agriculturists to get a clear understanding of public perception and sentiments about agriculture from various sources such as online forums, customer reviews, and social media platforms. Below are agricultural use cases where sentiment analysis can be utilized in digital agriculture:

- Agribusinesses can have valuable insight into consumer preferences, customer satisfaction, and market trends about a specific product's (fertilizer or pesticide) performance or their brands' product reputation.
- Farmers can make timely decisions by anticipating buying, selling, or storage price fluctuations of agricultural commodities.

- Sentiment analysis on public discussions helps advocacy groups to address misconceptions of sustainable agricultural practices.
- Government authorities can assess public support and people's opinions on newly proposed agricultural policies.
- Application of sentiment analysis through text classification strategy using supervised learning and handcrafted word lists can be applied to pest control management and commodity and food price prediction (Drury and Roche 2019; Bermeo-Almeida 2019; Huang and Zhang 2016).

The above discussions show that most data analytics approaches are based on Statistical, Data Mining, Machine Learning, and Natural Language Processing (NLP) techniques. Therefore, the next two sections have elaborated a brief overview of different Statistical, Data Mining, Machine Learning, and Natural Language Processing (NLP) techniques.

6.5 Statistical Approaches and Agricultural BigData Analytics

A few common types of statistical approaches employed for BigData analysis fall under the categories of Descriptive Statistics and Inferential Statistics.

6.5.1 Descriptive Statistics

Descriptive Statistics provides insight into data by summarizing the main characteristics of datasets using common measures of mean, median, mode, percentile, and standard deviation. The main features can be the dataset's Central Tendency, Variability, and Distribution.

Central Tendency refers to a statistical measure that is used to provide data insight with a single representative value of the dataset around which the data tends to cluster. The single representative value can be mean, median, or mode. In digital agriculture, central tendency value is used to calculate

- mean yield/acre or yield/hectare across multiple harvests or regions and helps farmers to access, record, and improve the overall performance of agricultural land or a particular crop
- mean concentrations of main soil nutrients i.e., Nitrogen (N), Phosphorus (P), and Potassium (K) in soil samples taken from different areas of a piece of agricultural land and helps farmers in identifying the areas of nutrient deficiency
- median prices of agricultural products in local markets provide insight into price fluctuations and market trends and ultimately assist farmers and agricultural businesses to optimize marketing strategies

- mean weights of livestock at different stages of growth that help ranchers to monitor growth rates, adjust feeding rates, and optimize their nutritional plan and breeding programs

Statistical Variability refers to the spread of data points (within a dataset) from each other and from the distribution center (any central tendency measures) and helps to get a clear understanding of diversity and range of values in the dataset. A few common measures of variability include range (calculated by subtraction of minimum value from the maximum value in a dataset), variance (the calculated average squared deviation of each data point from the mean value), and standard deviation (square root of variance). Within the context of digital agriculture, these measures can assist farmers in various ways i.e.,

- analysis of historical yield data along with the measurement of variability in soil characteristics and weather patterns assist farmers in making informed decisions about crop selection, planting, and harvesting schedules
- spatial and temporal variability measurement of soil nutrients and environmental factors enable farmers to improve crop performance and resource efficiency by optimizing the usage of water, fertilizer, and pesticides
- spatial and temporal variability measurement of pest populations and disease incidences help farmers to implement targeted pest management and disease control strategies.

Statistical Distribution (or probability distribution) describes the probability of occurrence of expected outcomes in a random experiment or observation. Statistical distribution is widely used in the field of agriculture in many ways i.e.,

- It enables agriculturists to predict the probability of obtaining different yield levels under varying environmental conditions
- Considering the distribution of soil properties, agriculturists can adapt optimal crop management practices i.e., crop rotation, irrigation scheduling, fertilizer application, etc.
- Analysis of the spatial and temporal distribution of pests help farmers to optimize the use of pesticide
- Climate modeling is based on probability distribution and helps agriculturists and policymakers to develop adaptation strategies to mitigate the impact of extreme weather events i.e., heavy rains, floods, heatwaves, etc.

6.5.2   Inferential Statistics

Based on sample data Inferential Statistical techniques are used to draw inferences about a population. Common techniques that fall under the category of inferential statistics include regression analysis, hypothesis testing, and confidence interval.

Regression Analysis technique is one of the widely used inferential statistical techniques that is applied in the agriculture sector to model a relationship between a dependent variable and one or more independent variables e.g.,

- crop yield can be predicted by regression analysis based on various factors i.e., soil nutrients (pH, nutrient level), crop management practices (irrigation, fertilization, pesticide application), and weather conditions (temperature, humidity, precipitation) that ultimately help agriculturists to optimize resource allocation.
- The relationship between crop growth parameters and environmental variables predicted by regression analysis can help to understand the crop's physiological responses to environmental factors (water availability, temperature, humidity, light intensity, etc.). This understanding helps agriculturists to make decisions about best crop management practices.
- Improvement in crop productivity is possible by considering regression models for the assessment of soil fertility level based on soil characteristics i.e., texture, nutrient concentrations, organic matter, etc.
- Optimized crop yield with minimum crop losses is possible by developing regression models that can predict the occurrence and severity of pest and disease outbreaks based on environmental factors (i.e., rainfall, temperature, humidity, etc.)
- The prediction of future prices of agricultural commodities based on geopolitical events, macroeconomic indicators, and supply and demand dynamics helps policymakers to make informed decisions about investment planning, marketing strategies, and crop production.

Hypothesis Testing techniques use sample data to make inferences about population parameters by formulating a Null Hypothesis (H_0) and assessing the strengths of evidence against this hypothesis by using statistical techniques on collected data. Below are some applications of hypothesis testing in agriculture.

- Agricultural scientists employ Hypothesis Testing methodology to assess the influence of soil management practices (including tillage, irrigation, and drainage) on soil characteristics like pH, nutrient levels, and water retention, as well as crop productivity. Through the collection of soil samples from experimental plots and subsequent analysis for various parameters, researchers can rigorously test hypotheses regarding the efficacy of different management strategies in enhancing soil health and optimizing crop performance.
- Agricultural scientists use Hypothesis Testing to conduct comparisons between plots treated with pest control measures and untreated ones to gauge the impact on pest incidence or severity. Using this approach, researchers ascertain whether the observed variations in pest pressure between the treated and untreated plots are statistically significant, thereby determining the efficacy of the treatment in mitigating pest issues.
- Hypothesis testing is one of the basic tools for evaluating the ecological impacts of farming techniques on soil erosion, water quality, greenhouse gas emissions, and biodiversity. Agricultural scientists perform comparative analyses across diverse

land management systems to analyze the implications of farming practices on the overall adaptability of ecosystems.
- Hypothesis testing plays a pivotal role in determining the significance of outcome differences of comparative trials that are frequently undertaken in agricultural research to estimate the effectiveness of diverse treatments such as the use of new crop varieties, fertilizers, pesticides, and cultivation techniques. For instance, agricultural scientists may use statistical tests to assess the significance of observed differences in crop yields with two varieties of a crop.

Confidence Interval is a statistical technique that provides a range of values based on sample data within which the actual population parameter (means, proportions, regression coefficients) is likely to lie. The use of confidence intervals in agricultural contexts has been explained below with the help of a few examples.

- Based on data collected from sample plots or farms, agricultural scientists can calculate confidence intervals for the overall average yield of the whole field or agricultural farm.
- Soil scientists with the help of calculated confidence intervals on soil samples collected from different locations of an agricultural field can estimate the average of overall soil properties (i.e., pH, organic matter content, nutrient level, etc.) in that agricultural field.
- By calculating the confidence interval on collected surveillance data from sample plots, entomologists can estimate the overall proportion of crops affected by diseases and pest attacks. Therefore, it indirectly helps entomologists to make informed decisions about pest or disease management strategies.
- Confidence interval helps agronomists to compare the effectiveness of agricultural inputs (i.e., irrigation, fertilizer, pesticide, etc.) that are used in different proportions in field trials and experiments.
- The analysis of data collected from sampling stations supports the calculation of confidence intervals for various environmental parameters such as greenhouse gas emissions and air pollutant concentrations that ultimately help agriculturists to estimate environmental risks and develop sustainable farming practices.

6.6 Role of Machine Learning and NLP for Agricultural BigData Analytics

Machine learning algorithms based on data-driven methodology exploiting past experiences (stored in different data formats) are able to generate and reconstruct knowledge schemes for the correct future predictions. The typical working of machine learning applications can be divided into two phases i.e., the training phase and the testing phase. The training/learning phase can utilize (binary, numeric, ordinal, or nominal) feature vectors as input vectors to perform a task from experience. After deducing rules (mathematical or statistical relationships), the trained model is used for the identification, classification, and clustering of data. During the testing phase, unknown samples are provided to the trained model to check the accuracy of the trained model (Benos et al. 2021). Followings are different types of machine learning schemes i.e.,

- Supervised Learning: finding the optimal output on the basis of known input and output labels.
- Unsupervised Learning: generate structures through given inputs in the absence of labelled data.
- Reinforcement Learning: where intelligent agents based on the trial-and-error method take actions to maximize the positive outcome.
- Deep Learning: a subset of machine learning algorithms based on the concept of multi-layered Artificial Neural Networks (ANNs) mimicking the processing of the human brain used to approximate predictions with high accuracy. Deep learning algorithms with processing at multiple layers learn data representations at multiple levels of abstraction (LeCun et al. 2015). One of the key differences between deep learning and machine learning approaches is that deep learning demands more data than machine learning for accurate classification.

The use of these machine learning and deep learning approaches in agricultural contexts has been elaborated below (Liakos et al. 2018).

- Association rule mining approaches (i.e., unsupervised Machine Learning Apriori and FP-Growth Algorithms) assist in the extraction of meaningful patterns and relationships from agricultural datasets. By analyzing vast datasets encompassing agricultural variables (i.e., crop types, weather conditions, soil characteristics, and farming practices), association rule mining uncovers hidden correlations and dependencies. For example, these approaches can identify associations between specific weather patterns and crop yields or reveal patterns in pest occurrences concerning environmental factors.
- Clustering algorithms (i.e., unsupervised Machine Learning based K-mean clustering, hierarchical clustering, etc.) help agriculturists to uncover inherent groupings within available datasets. For example, using agricultural datasets (containing variables of soil composition, pest infestation rate, crop yield, etc.) clustering facilitates the identification of distinct clusters or categories of agricultural phenomena i.e., soil types with similar nutrient compositions, high-yield zones, or regions susceptible to specific pests. Ultimately, the clustering approach empowers agricultural stakeholders to make informed decisions, optimize resource allocation, and enhance overall productivity and sustainability in farming practices.
- Classification empowers agricultural stakeholders to proactively address the challenges of disease detection, crop yield prediction, etc. by recognizing patterns and making predictions of unseen data points through training of classification models (i.e., supervised machine learning based Decision Trees, Support Vector Machines (SVM), K-Nearest Neighbours (KNN), Naïve Bayes, Neural Network algorithms, etc.) on historical agricultural data. The proactive measures taken by agriculturists enhance overall agricultural productivity and environmental sustainability.

- Deep Learning approach leveraging on the use of ANNs with multiple layers to extract high-level features from large volumes of agricultural data (including images, sensor readings, environmental variables, etc.) enables farmers to improve agricultural productivity by making informed decisions. For example, Convolutional Neural Networks (CNNs) are adept at analyzing agricultural images to monitor crop growth stages, assess plant health, and detect crop diseases.

In summary, the generic domains of agriculture exploiting machine learning approaches include.

- Crop Management where machine/deep learning algorithms are used to predict crop yield and quality.
- Identification and classification of crops, plants, and fruits.
- Identification and classification of crop diseases, weeds, and pest attacks in agricultural fields.
- Irrigation management where machine/deep learning solutions provide optimal use of water resources.
- Detection and identification of illness signs in livestock by tracking data on animal's vital signs, movement patterns, and feeding behaviours.
- Prediction of demand fluctuations for agricultural products in the market.
- Optimization of supply-chain operations such as delivery routes and inventory levels.

Natural Language Processing (NLP) has emerged as a vital component of text mining which is an interdisciplinary subfield of computer science and information retrieval. NLP enables computing systems to analyze, comprehend, generate, and manipulate natural (or human) language text or voice data. Through the use of NLP, farmers' communities can learn better in their local language from their peers and public/private agricultural organizations located in different regions. Therefore, currently, there is a trend in the researchers' community to enable digital agriculture services through the development of NLP-based agricultural decision support systems. Table 6.4 describes a few usages of NLP in smart agriculture.

Agrobot (Gounder et al. 2021) and ADSS (Prasad et al. 2008) are examples of smart farm systems that use NLP.

6.7   Data Visualization in Digital Agriculture

Data visualization can be considered as a complementary technique that is used in conjunction with data analytics to graphically present discovered patterns and data insights in an understandable way (Sadiku et al. 2016). In recent times, agriculture is one of the data-intensive industries that involve many factors i.e., soil, crop, weather, diseases, pests, policies, market trends, etc. Therefore, in digital agriculture, data visualization assists agricultural stakeholders (including farmers, policymakers, researchers, pesticide/fertilizer companies, extension workers, consumers, etc.) with

Table 6.4 Usage of NLP in digital agriculture

Agricultural unit	Type of text data to analyze	Benefit
Crop monitoring and pest management	• Agricultural literature • Pest databases • Online forums	NLP techniques by detecting information about pest attacks, crop diseases, and appropriate control measures helps agriculturists with real-time alerts about early signs of crop threats and appropriate recommendations
Customer feedback analysis	• Social media comments • Customer reviews • Feedback surveys	NLP models by extracting valuable insights into customer perceptions helps agribusinesses to improve product quality
Market intelligence and price forecasting	• News articles • Social media platforms • Market reports	Assists to forecast price fluctuations and predict market trends
Research and innovation	• Scientific literature • Research papers and patents	NLP approaches by identifying research trends facilitate knowledge discovery
Regulatory compliance and policy analysis	• Regulatory documents Government policies • Industry standards	Helps agricultural organizations to stay informed about regulatory compliance

intuitive visualizations to make informed decisions (Liu 2021; Kumari et al. 2024; Kavitha et al. 2024). Different stakeholders have different perspectives regarding the creation and usage of data visualizations as described in Table 6.5.

Table 6.5 Usage perspectives of data visualization by agricultural stakeholders

Agricultural stakeholder	Visualization usage perspective(s)
Farmers	Cost reduction, optimize the usage of farm resources, optimize production, increase profits
Researchers	Hypothesis testing, new knowledge discovery, model validations
Policy Makers	Impact evaluation of policies
Pesticide/Fertilizer Companies	Improve product outcome, improve production, increase profit
Extension Workers	Improved farmer counseling for managing their land, crops, and livestock
Policymakers	Designing and evaluation of new policies
Consumers	A better understanding of quality and sustainability by viewing the origin, nutritional value, and traceability of agricultural products

Concerning the effectiveness of visualizations in digital agriculture, it is important to consider best practices for the creation of data visualizations for agriculture, for example

- Define the purpose of visualization by considering the demands of the concerned audience
- Select the right data (relevant, accurate, and complete) and the right format (appropriate, clear, and engaging). For instance, bar charts, pie charts, line charts, and maps can be used to show comparisons among categories, proportions of a whole, trends over time, and spatial distribution, respectively.
- Use consistent design elements i.e., font, font sizes, colors, etc. to avoid audience confusion and appropriate legends, axes, and scales representing correct meanings, direction, and magnitude, respectively.
- Include interactive features such as filters and zooms to allow the audience to customize and explore data easily.

A few examples of effective data visualizations generated using interactive charts available on the online platform [worldindata.org (https://ourworldindata. org/)] demonstrating different records of the agriculture sector have been shown in Figs. 6.2, 6.3 and 6.4. It can be clearly seen that these interactive charts in different forms (i.e., choropleth maps, line graphs, and stacked area charts) with various user-friendly controls (i.e., selection tool, filtering, panning, zooming, tooltips, etc.) enable users to get deeper insights by dynamic customization of the visualizations.

Questions

Q6.1: What are the primary, secondary, and tertiary data sources utilized in agricultural data analytics?

Q6.2: How does data analytics contribute to enhancing crop quality and yield in the context of smart farming?

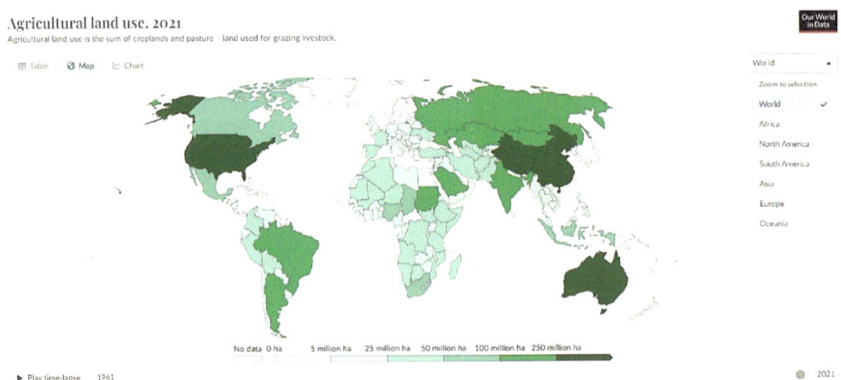

Fig. 6.2 Interactive choropleth maps visualization depicting world's agricultural land usage (1961–2023)

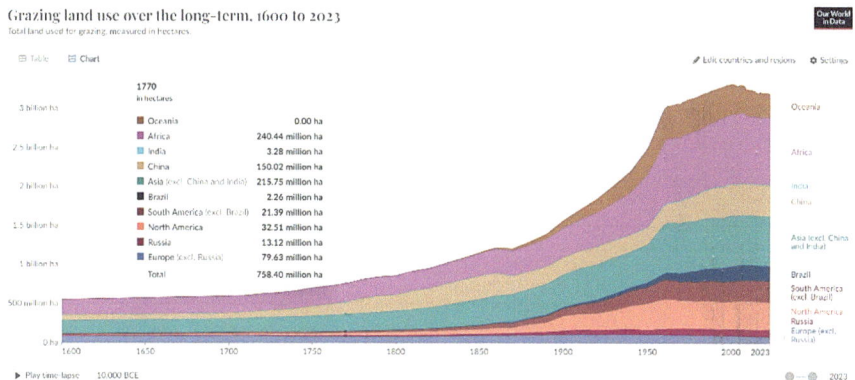

Fig. 6.3 Interactive stacked area chart visualization depicting world's long-term grazing land usage (1600–2023)

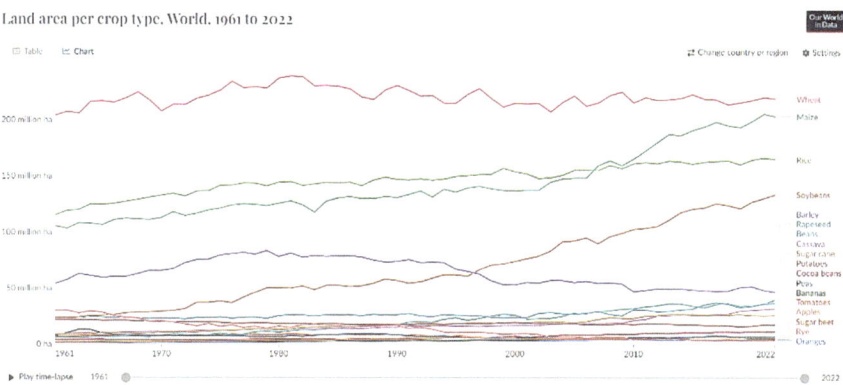

Fig. 6.4 Interactive line graphs visualization depicting land area/crop (1961–2022)

Q6.3: Which machine learning algorithms can be applied in agricultural data analytics to automate pest and disease classification in crops?

Q6.4: What kind of interactive chart visualization would you prefer to show the production of different crops in different countries and years? Which type of user-friendly control you would like to include as part of interactive chart visualization?

Q6.5: Explore and write the name of a few data analytics applications developed for livestock management, specifically for predictive health monitoring and feed optimization.

References

Allahyari M et al (2017) A brief survey of text mining: classification, clustering and extraction techniques. arXiv preprint arXiv:1707.02919

Arsevska E (2017) Elaboration of a semi-automatic method for identification and analysis of signals of emergence of animal infectious diseases at international level

Arsevska E et al (2018) Web monitoring of emerging animal infectious diseases integrated in the French animal health epidemic intelligence system. PLoS ONE 13(8):e0199960

Benos L et al (2021) Machine learning in agriculture: a comprehensive updated review. Sensors 21(11):3758

Bermeo-Almeida O et al (2019) Sentiment analysis in social networks for agricultural pests. In: 2nd International conference on ICTs in agronomy and environment. Springer

Chakraborty S et al (2016) Predicting socio-economic indicators using news events. In: Proceedings of the 22nd ACM SIGKDD international conference on knowledge discovery and data mining

De Quincey E, Kostkova P (2010) Early warning and outbreak detection using social networking websites: the potential of twitter. In: Electronic healthcare: second international ICST conference, eHealth 2009, Istanbul, Turkey, 23–15 Sept 2009, Revised Selected Papers 2. Springer

Dhar V (2013) Data science and prediction. Commun ACM 56(12):64–73

Drury B, Roche M (2019) A survey of the applications of text mining for agriculture. Comput Electron Agric 163:104864

Drury B et al (2016) The extraction from news stories a causal topic centred Bayesian graph for sugarcane. In: Proceedings of the 20th international database engineering & applications symposium

Feldman R, Sanger J (2007) The text mining handbook: advanced approaches in analyzing unstructured data. Cambridge university press

Goel R et al (2018) EpidNews: an epidemiological news explorer for monitoring animal diseases. In: Proceedings of the 11th international symposium on visual information communication and interaction

Gounder S et al (2021) Agrobot: an agricultural advancement to enable smart farm services using NLP. J Emerg Technol Innov Res

Han J, Pei J, Tong H (2022) Data mining: concepts and techniques. Morgan Kaufmann

Huang R, Zhang W (2016) Study on sentiment analyzing of internet commodities review based on word2vec. Comput Sci 43(s1):387–389

Issad HA, Aoudjit R, Rodrigues JJ (2019) A comprehensive review of data mining techniques in smart agriculture. Eng Agric Environ Food 12(4):511–525

Kamilaris A, Kartakoullis A, Prenafeta-Boldú FX (2017) A review on the practice of big data analysis in agriculture. Comput Electron Agric 143:23–37

624 J et al (2024) Role of data visualization and big data analytics in smart agriculture. In: AI Applications for business, medical, and agricultural sustainability. IGI Global, p 160–196.

Kim J, Cha M, Lee JG (2017) Nowcasting commodity prices using social media. PeerJ Comput Sci 3:e126

Küker S et al (2018) The value of necropsy reports for animal health surveillance. BMC Vet Res 14:1–12

Kumar A, Biswas P, Sharan A (2016) Identifying crop specific named entities from agriculture domain using semantic vector. Information systems design and intelligent applications. Springer, pp 595–603

Kumari S, Pandey, Tiwari S (2024) data visualization techniques in smart agriculture implementation. In: AI applications for business, medical, and agricultural sustainability. IGI Global, pp 122–159

LeCun Y, Bengio Y, Hinton G (2015) Deep learning. Nature 521(7553):436–444

Liakos KG et al (2018) Machine learning in agriculture: a review. Sensors 18(8):2674

Liao W-T et al (2015) Improving farm management optimization: application of text data analysis and semantic networks. In: 2015 ASABE annual international meeting. American Society of Agricultural and Biological Engineers

Liao W-T et al (2016) From Tweets to farm management: mining agricultural information and discovering agricultural communities in social networks. In: 2016 ASABE annual international meeting. American Society of Agricultural and Biological Engineers

Liu W (2021) Application of data visualization and big data analysis in intelligent agriculture. J Comput Inf Technol 29(4):251–263

Liu B, Zhang L (2012) A survey of opinion mining and sentiment analysis. Mining text data. Springer, pp 415–463

Liu Y et al (2020) From industry 4.0 to agriculture 4.0: current status, enabling technologies, and research challenges. IEEE Trans Industr Inform

Malarkodi C, Lex E, Devi SL (2016) Named entity recognition for the agricultural domain. Res Comput Sci 117(1):121–132

Manning C, Raghavan P, Schütze H (2010) Introduction to information retrieval. Nat Lang Eng 16(1):100–103

Okurowski ME (1993) Information extraction overview. National Computer Security Center Fort George G Meade Md.

Prasad J, Prasad R, Kulkarni U (2008) A decision support system for agriculture using natural language processing (ADSS). In: Proceedings of the international multiconference of engineers and computer scientists. Citeseer

Sadiku M et al (2016) Data visualization. Int J Eng Res Adv Technol (IJERAT) 2(12):11–16

Vassakis K, Petrakis E, Kopanakis I (2018) Big data analytics: applications, prospects and challenges. In: Mobile big data: a roadmap from models to technologies, pp. 3–20

Weersink A et al (2018) Opportunities and challenges for big data in agricultural and environmental analysis. Annu Rev Resour Econ 10:19–37

Wolfert S et al (2017) Big data in smart farming–a review. Agric Syst 153:69–80

World Data. https://ourworldindata.org/

Ye N (2003) The handbook of data mining. CRC Press

Chapter 7
Impacts and Future Directions of Digital Agriculture

7.1 Learning Objectives

After studying this chapter, students will be able to

- describe the role of human–computer interaction and web services in the development of agricultural expert and decision support systems
- compare between agricultural expert and decision support systems
- explain the benefits of Digital Agriculture along with the future directions in this domain.

7.2 Impacts of Digital Agriculture

The transformation of agriculture from traditional to digital agriculture has reshaped the landscape of conventional farming with the infusion of digital technologies. From field data acquisition (through sensors) to the presentation of advanced data analytics (using AI techniques), the involvement of digital technologies has revolutionized every aspect related to the management of agricultural operations. The management of ongoing agricultural operations and activities (at outdoor farmlands, indoor green-houses, vertical farms, hydroponic/aquaponic systems, etc.) is complex and is characterized by substantial degrees of uncertainty due to the involvement of several environmental factors. Therefore, farmers are required to monitor various parameters (i.e., soil nutrient(s) level, humidity, sunlight intensity, rainfall, water availability, weather conditions, climate change, etc.) and ongoing activities (sowing, plowing, irrigation, fertilization, etc.) at farms. Digital agriculture is taking advantage of various digital technologies to automate the processes of field monitoring and management that ultimately help farmers with in-time decision-making about the optimal utilization of farm resources. The optimal utilization of resources (in terms of water, fertilization, fertigation, pesticide spray, etc.) through various enabling digital technologies

SpringerBriefs in Agriculture, https://doi.org/10.1007/978-3-031-67679-6_7

is not only helpful to obtain high yield but also promotes environmentally sustainable farming practices. Previous chapters of this book provide details regarding the role of various enabling digital technologies that pave the way for the realization of digital agricultural systems. In recent times, several agricultural expert systems and agricultural decision support systems have been developed based on the concepts and technologies related to BigData storage and BigData analytics (Talari et al. 2022). Most of these agricultural expert systems and agricultural decision support systems are web-based interactive systems and the successful development of these systems is heavily reliant on principles of human–computer interaction (HCI) and web technologies (Fisher et al. 2012). Therefore, before mentioning details of agricultural expert and decision support systems, a brief overview of the principles of HCI and web technologies is part of Sects. 7.3 and 7.4. In the end, this chapter describes the profound impacts of digital agriculture in terms of multifaceted benefits that resonate across the entire food chain of the agricultural ecosystem through sustainable and resource-efficient practices.

7.3 Human–Computer Interaction (HCI)

HCI is a multi-disciplinary field of study that focuses on designing digital interfaces to ultimately enhance the user experience with computing devices. Thus, HCI can be considered a subject area that highlights the involvement of users and technology in the design process (Dix et al. 2000; Saizmaa and Kim 2008). HCI researchers observe human behavior while interacting with heterogeneous computing devices and design technologies that allow humans to interact with computers in easy and user-friendly ways. Modern agriculture is highly technical as a variety of agricultural stakeholders i.e., farmers, ranchers, extension workers, researchers, advisers, and policymakers are concerned with getting feedback from Agricultural decision support systems. Therefore, as the exposure level varies for different agricultural stakeholders, consequently the agricultural domain highly demands the development of efficient HCI interfaces regarding user-centered design practices. In HCI, user-centered design practices deal with the development of computing systems while focusing on the needs and limitations of the users (Marques 2017). Concerning HCI research, the agriculture domain is an underserved area (Posadas et al. 2021) and it is informed in literature studies that one of the reasons for farmers' reluctance to use computer-based support systems is the designing complexity of agricultural decision support systems. On the other hand, it is also observed that digital agricultural systems with simple interfaces while taking into account the demands of particular stakeholders are always preferred over conventional approaches (Rose et al. 2018). Therefore, concerning the user-centered design of agricultural decision support systems, the continuous involvement of agricultural stakeholders through a feedback mechanism is a very important aspect. Several agricultural applications based on user-centered design approaches have been developed i.e.,

- farmer-computer interaction systems,
- crowdsourcing precision agriculture applications, and
- various types of map visualization agricultural systems (to show crop production, weather forecast, pest and disease control, irrigation management, dairy farming, etc.) have been developed while taking into account the diverse needs of the agriculturists and the context of use (Marques 2017; Posadas et al. 2021; Gutiérrez et al. 2019).

7.4 Web Technology and Web Service

Web applications (websites) are a collection of interlinked web pages hosted on different web servers and are accessible in web browsers through a unique address known as a Uniform Resource Locator (URL) i.e., www.website.com. Websites are developed for human consumption and are of two types i.e., static websites and dynamic websites. Static websites are simple web pages developed using simple languages i.e., Hypertext Markup Language (HTML), Cascaded Style Sheets (CSS), etc. where contents remain the same. Static websites are not connected to databases. Dynamic websites represent the collection of web pages where contents change dynamically for each user against each request. The web pages in dynamic websites are developed using different server-side scripting languages i.e., Java Server Pages, Personal Home Page (PHP), Active Server Pages (ASP) (Ahmed 2013; Nixon 2012), etc., and connected to backend databases. Several web applications have been developed for agricultural farming systems e.g.,

- Blue Harvest Farms (https://www.blueharvestfarms.com/)
- Monte Vista Farming (https://www.montevistafarming.com/)
- Harvest Valley Farms (https://www.harvestvalleyfarms.com/)
- North Star Orchard (https://northstarorchard.com/).

Contradictory to web applications, web services are developed as software systems or components to be consumed or used by computers or machines to interact with each other over the Internet or intranet (Alonso et al. 2004; Papazoglou 2008; Roy and Ramanujan 2001). Web services can be developed using Simple Object Access Protocol (SOAP) or Representational State Transfer (REST) standards. Concerning certain associated advantages (i.e., ease of coding, caching, low bandwidth requirements, etc.), RESTful web services are more popular than SOAP web services (Halili and Ramadani 2018). CLUeFARM (Colezea et al. 2018) is one of the examples of an integrated web service platform for smart farms.

7.5   Agricultural Expert Systems and Decision Support Systems

An Expert System is an automated computing system that is developed to reproduce the performance of (specific) domain expert(s) to solve a real-world problem. Expert systems may or may not have learning components but are implemented either to assist human workers (through some information system) or to solve real-world problems. The core components of Expert Systems include the *knowledge base:* collection of (production) rules and related information structures derived from a human expert, *Inference engine:* to acquire and interpret related data from the knowledge base for finding a solution to users' problems, *Learning Module:* for getting more pertinent data from various sources, *User Interface:* to present questions/information (in the form of images, video clips, animation clips) to users to get user responses to forward it to the inference engine. Expert systems in the agriculture domain are especially helpful to farmers in solving the problem complexities related to crop selection, soil erosion, pesticide cost, pest resistance, yield losses, diminishing market prices, etc. These agricultural systems are advisory systems that assist integrated decision support systems for crop/livestock production and management. Implementing these expert systems is important, as a farmer may not become proficient in managing all aspects of farming practices. These expert systems are developed based on the experience and knowledge of agricultural scientists to assist farmers by providing advisory support (Maurya et al. 2013; Sriram and Philip 2016; Akankasha and Vikas 2014). A few examples of agricultural expert systems are:

(1) GRAPE (Saunders et al. 1987) expert system provides recommendations to grape growers related to insect/pest management.
(2) Expert System for Irrigation Management (ESIM) (Srinivasan et al. 1991) helps the farmer in making decisions on water management.
(3) CROPLOT (a rule-based expert system) (Nevo and Amir 1991) was proposed to determine the suitability of crops to a specific plot/land area.
(4) COMAX (Lemmon 1990) provides information to farmers and managers related to integrated crop management in a cotton field.
(5) CITEX expert system (Salah et al. 1993) was developed to provide various services i.e., on-farm assessment, irrigation, and fertilization scheduling of orange orchards.
(6) LIMEX expert system (Mahmoud et al. 1995) assists lime growers with multimedia to help them with cultivation time.
(7) The expert system POMME (Roach et al. 1985) helps with disease and insect-pest management activities in the Apple orchard.
(8) A multilingual expert system called VEGES (Yialouris et al. 1997) was developed for the diagnosis of insects/pests, diseases, and nutritional disorders of various vegetables (i.e., tomato, cucumber, pepper, lettuce, etc.) in greenhouses.
(9) AMRAPALIKA expert system (Prasad et al. 2006) helps in the diagnosis of pests, diseases, and disorders in Indian mangoes.

(10) CALEX (Plant et al. 1989) is helpful in the diagnosis of peach and nectarine disorders.
(11) Distance Diagnostic and Identification System (DDIS) (Sprenkel and Momol 2002) facilitates the diagnosis and identification of plants' diseases and pest attacks.
(12) To control various diseases of Sugarcane, the D-CAS expert system (Krit and Baudin 1994) was implemented.

Other than Expert Systems, interactive software-based systems known as Decision Support Systems (DSS) have been proposed to enable decision-makers to better understand processes and unexpected situations while taking into account complex factors. Cost/time savings, improved communication, and better use of data/resources are other associated advantages of DSS. The fundamental difference between DSS and ES is that a DSS is an interactive software system that facilitates decision-making while utilizing complex data in an unstructured environment. On the other hand, ES are computer programs that automate decision-making in a specialized problem domain to replace the human decision-maker(s) (Ford 1985; Finlay 1990). Different types of DSS systems have been proposed and implemented under different categories of agriculture domains as shown in Table 7.1 (Zhai et al. 2020).

Table 7.1 Agricultural decision support systems

Agriculture domain	DSS	Purpose
Mission planning	• AgriSupport II system (Recio et al. 2003) • Multi-robot sense-act system (Conesa-Muñoz et al. 2016) • Route planning in soil-sensitive system (Bochtis et al. 2012)	To provide guidance for farm operation scheduling, herbicide treatment, and route planning for an agricultural vehicle.
Water resources management	• SIDSS (Navarro-Hellín et al. 2016) • FDSS (Giusti and Marsili-Libelli 2015)	To assist (through advice) on irrigation scheduling and reports.
Climate change adaptation	• LandCaRe DSS (Wenkel et al. 2013) • GIS-based DSS (Kadiyala et al. 2015)	To help farmers with farm management and increase crop productivity under climate change.
Food waste control	• Quality sustainability DSS (Ting et al. 2014) • DSS for e-grocery deliveries (Fikar 2018) • Beef supply chain (Soysal et al. 2014)	To support delivery plans for transporting e-grocery, wine, and beef.

7.6   Benefits of Digital Agriculture

Advantages associated with the realization of digital agriculture for both local-scale farming and commercial-scale farming (Elijah et al. 2018) (shown in Fig. 7.1) have been described below.

7.6.1   Increased Production

Automation of agricultural operations (i.e., timely plowing, sowing, breeding, cultivation, planting, watering, etc.) and optimized application of fertilizer, water, and pesticide to crops eventually increase the farm production rate.

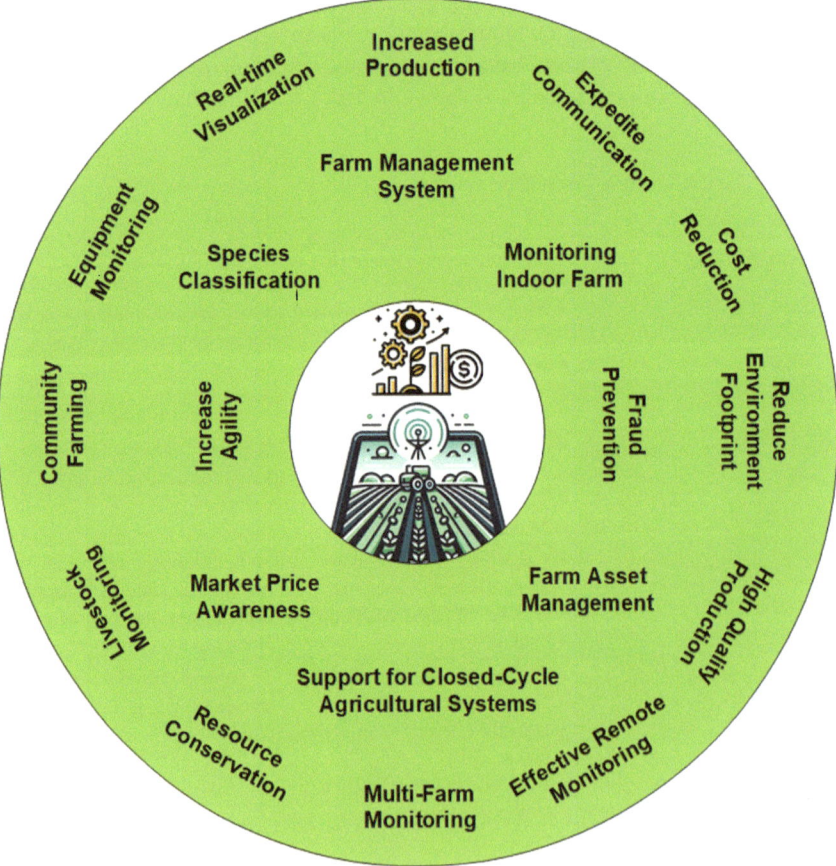

Fig. 7.1 Advantages of digital agriculture

7.6.2 Expedite Communication

Transmission of collected agricultural data (i.e., field data, environment data, yield data, storage data, etc.) is quick and easy through the use of advanced wireless technologies and digital (mobile) devices as compared to manual distribution (using paper forms).

7.6.3 Effective Remote Monitoring

Sensing technology along with drone technology integrated with systems having image processing capabilities allow farmers to remotely monitor field conditions, soil quality, sowing patterns, irrigation needs, weed activity, and pest populations at their outdoor farmlands and indoor greenhouses. Moreover, farmers with the help of IoT devices can simultaneously monitor the ongoing activities at their multiple fields that are located at different geographical locations). The remote monitoring of crops, plants, livestock, land, and soil on multiple farms assists farmers in real-time decision-making.

7.6.4 Real-Time Visualization

Digital farming helps farmers to visualize different environmental factors (i.e., light intensity, humidity, temperature, soil moisture, etc.) and farm productivity in real-time to assist the decision-making process.

7.6.5 Community Farming

In rural areas, digital agriculture through the use of mobile apps, IoT equipment, and IoT services (free or paid) helps farmer communities to have ubiquitous interaction with agriculture experts by sharing field data and information stored on a common data storage repository.

7.6.6 Automation of Agricultural Farm Management System

The adoption of digital technologies (i.e., sensing technologies, communication and networking technologies, Global Positioning System (GPS), drones, artificial intelligence, etc.) in agriculture supports the realization of automatic farm management systems. The automation at farmland includes the automation of various activities i.e.,

- automatic sowing and cultivation of crops through automatic tractors
- automatic inspection or monitoring of farmland or greenhouse environmental factors and ongoing farm activities
- automatic irrigation management
- automatic pest and animal protection
- automatic disease detection and prevention
- automatic fertilization
- automatic spraying of pesticides and herbicide
- automatic harvesting
- automatic monitoring of animal behavior (feed time) and tracking etc.

that ultimately increases crop yield and livestock productivity.

7.6.7 High-Quality Production

The way of optimized treatments (watering, pesticide, insecticide, fertilizer, etc.) at each phase of crop development from sowing to harvesting increases agricultural yield and its quality. Moreover, product quality analysis and its correlation to given treatment are helpful for farmers to adjust parameters to obtain a high-quality product in the future.

7.6.8 Safety and Fraud Prevention

Other than the challenges of higher agricultural production, the safe supply of these agricultural products without fraud (i.e., counterfeit, adulteration, artificial enhancement, etc.) to end-users is also necessary. Digital agriculture through the integration of sensing technology and blockchain-based smart contract technology ensures product integrity, process integrity, product quality, and logistics traceability of the agri-food supply chain. Smart contracts (also known as distributed ledger technology) with the characteristics of immutable transaction history of raw agriculture product material to end-user food packaging solve the problem of food safety and quality.

7.6.9 Agility

Digital agriculture increases the farmer's agility to respond to any change in environmental conditions that can significantly affect crop production, livestock health, field soil quality, etc.

7.6.10   Equipment Monitoring

Taking into account different factors (i.e., production rates and labor effectiveness), all types of farm equipment (i.e., seeding/planting equipment, hay/forage equipment, spray equipment, loaders, tractors, etc.) can be monitored through various digital devices.

7.6.11   Security and Management of Farm Assets

Digital agriculture with the assistance of IoT devices promotes real-time security by monitoring of farm equipment and enables the in-time replacement of machinery, parts, equipment theft, and timely routine maintenance.

7.6.12   Livestock Monitoring and Management

The outfitting of digital devices (with concerned technologies i.e., sensor technologies, GPS, wireless communication, data management) to animal's ears, collar, body, legs, tails, udders, etc., and ingestible capsules with microchips have been used in livestock farming to assist farmers for the

- animal identification
- rumination (duration) monitoring
- tracking and monitoring of herd movement and overall wellness
- tracking of animal activities within the farm
- animal health monitoring (including heat/cold stress)
- detection of heat to improve the conception rate
- detection of the best time to inseminate
- live monitoring of calving
- record-keeping of feed, medication, treatments, and outputs.

7.6.13   Support Closed-Cycle Agricultural Systems

Plantation and growing of agri-food in smart closed-cycle agricultural systems (i.e., hydroponic, aquaponic, vertical farming, on skyscrapers' rooftops and walls, etc.) become a reality due to the use of digital technology in agriculture. These IoT-based smart closed-cycle agricultural systems benefit the urban population with fresh food supplies.

7.6.14 Market Price Awareness and Profit Creation

The ubiquitous availability of smartphones and low-cost mobile applications, network services, digital devices, etc. helps farmers to get updated information about market prices. This kind of awareness about market prices of agricultural products also assists farmers in building direct relationships with consumers by avoiding the exploitation of brokers or middlemen. On the other side, consumers can easily locate and directly contact farmers for fresh agricultural products.

7.6.15 Cost Reduction

Optimized treatment i.e., accurate watering, fertilization, pesticide spray, in-time disease management, etc. at each stage of crop cultivation (from sowing to harvesting) directly affects the cost of production.

7.6.16 Decision-Making Efficiency

Both the government and non-governmental organizations (which are related to the agriculture sector) get benefits from collected data gathered by IoT devices implemented or deployed in outdoor farmlands or indoor greenhouses. Concerning farmers, the collected data serves the purpose of accurate and timely decision-making for various farm management processes and activities. On the other hand, the organizations used collected data for large-scale treatments and agriculture interventions i.e., disease spread, veld fire outbreaks, etc.

7.6.17 Recognition and Classification of Plant Species

Digital technologies assist in the recognition of different plants, vegetables, and fruits at different stages of their growth, which ultimately is helpful for automatic harvesting purposes.

7.6.18 Monitoring of Indoor Farming

The use of digital technologies (i.e., low-cost, low-power sensors, wireless communication, drones, digital cameras, etc.) within the closed-cycle indoor farms (e.g., greenhouses, vertical farms, hydroponic, aquaponic farms, etc.) reduces the cost of manpower related to the remote monitoring and management of different on-going activities.

7.6.19 Sustainable and Resource-Efficient Practice

Besides the primary benefits (discussed in the above section), digital agriculture is effectively contributing to environmental sustainability as it helps to preserve crucial resources to sustain a better lifestyle for future generations. For example, optimized usage of resources (i.e., water, fertilizer, pesticide spray, energy, land, etc.) helps farmers to conserve resources accurately by allocating these resources as needed within the farm. Below, it has been discussed how resource conservation through digital technology ultimately affects the ecological and environmental footprint positively.

Water Saving in Land Irrigation

Different areas of land need different levels of water for irrigation. The digital agricultural system with integrated sensors and GPS systems can detect the location of the needy land to apply precise pressure. The agricultural sensors identify the exact amount of water required for a piece of land and this practice not only saves water but also protects the crops from damage by over-supply of water.

Climate Change

Traditional farming practices caused the emission of catastrophic greenhouse gases. The digital agricultural revolution reduces the predictable calamities of climate change and guarantees a safer environment for living beings. Moreover, it minimizes the CO_2 emissions in the air to provide a cleaner living environment for everyone.

Better Resource Utilization

Besides water saving in land irrigation, digital agriculture reduces the cost of fertilizers and chemicals by avoiding unnecessary usage and applying them where the soil needs them (Rose et al. 2018). The sensors check the availability of land before applying accurate agricultural procedures on them. This sensor information is used to avoid duplication of sprays, fertilizers, pesticides, or water to ensure the availability of resources in the long run. These smart agricultural choices are not only efficient to save money but are also beneficial for the ecosystem.

Marine Ecosystems Protection

In the past, agricultural practices damaged marine ecosystems. The harmful chemicals and waste thrown in the rivers polluted the water and caused a major threat to marine life. The use of advanced technologies in agriculture prevents toxic fuels from being disposed of in rivers and seas by reducing the fuels and chemicals. The less fuel disposed of in the sea, the better it is for the marine ecosystems.

Flora and Fauna Sustainability

Excessive pesticides are harmful to wildlife and wild plant species (Fisher et al. 2012). Better and revolutionized digital agricultural practices (i.e., smart irrigation, fewer pesticides on weeds, habitat mapping using GIS, sustainable land use, early disease detection, etc.) protect nature and animals (especially the endangered species). Moreover, the use of agricultural devices and robots to complete most of the agricultural operations ensures the preservation of biodiversity by providing habitat and refuge for native flora and fauna.

Improvement in Grassland Productivity and Sustainability

Digital technologies assist in monitoring animals that are directly associated with grassland productivity. For example, the finer monitoring and control of grazing animals ultimately increases resource efficiency such as better distribution of nutrients and reducing the need for fertilizer in the grassland. A better understanding of soil and pasture resources helps in the prevention of under- or over-utilization of forage at specific sites. Moreover, measuring animals' urine excretions assists in developing strategies to mitigate nitrogen loss on farmland.

Minimize Salinity

Soil salinity is one of the most vicious environmental factors that not only limits the productivity of crops but is a constant threat to the ecology and agriculture landscape. Salinity is threatening as most of the crops are sensitive to high concentrations of salts in the soil and the affected area is increasing day by day. Digital agriculture techniques and methods i.e., remote sensing and salinity sensors have provided promising services to detect soil salinity with the ultimate aim to minimize soil salinization and maximize crop productivity. Moreover, for soil salinity mapping, IoT-based salinity mapping schemes at the irrigation stage are more accurate and cost-effective way compared to the standard chemical analysis approach.

7.7   Future Directions in Digital Agriculture

Supported by the use of advanced ICT, digital agriculture is the need of the hour to automate the traditional practices of agriculture to get high productivity at a lower cost. Considering the continuous advancements in technology, the future of digital agriculture holds several exciting possibilities, for example,

- Digital twin technologies will be able to create more sophisticated replicas of physical farms that ultimately help farmers in decision-making (for farm resource allocation and crop management) by simulating various scenarios.
- The effectiveness and scalability of digital agriculture applications will be enhanced with the availability of faster and more reliable (5G) communication systems.

- Seamless integration of diverse data sources will enable enhanced data sharing and improved interoperability.
- Global adoption of digital technologies in agriculture will enhance knowledge sharing and global collaboration to address major challenges in agriculture in different parts of the world.
- The widespread adoption of blockchain technology in agriculture will improve the traceability (information about origin and journey) of agricultural products that are part of the food supply chain.
- More precise monitoring of crops, livestock, and environmental conditions is expected through the development of more intelligent sensors.
- Enhanced personalized recommendations to farmers will not only help to improve crop and livestock yield but assist agriculturists to optimize and reduce the usage of farm resources and environmental impact, respectively.
- Data collection through crowdsourcing (involving diverse farming communities) enhances the accuracy of agricultural data analysis.
- Data processing closer to its source (edge computing) will reduce latency and be helpful in real-time decision-making in farms.
- Innovations in (hyperspectral and multispectral) imaging technologies will improve the accuracy of early plant disease detection and pest attacks.
- The adoption of robots to perform agricultural operations (i.e., plowing, sowing, planting, weeding, harvesting, etc.) will reduce labor costs with improved overall crop and livestock productivity.
- Greater utilization of advanced AI algorithms will improve prediction accuracy and personalized farming strategies.
- Modern digital agriculture approaches will promote regenerative farming practices (rotational grazing, mixed crop grazing, cover cropping, etc.) to minimize environmental impact and improve sustainability.
- The integration of farm management software will foster a holistic approach to digital agriculture by offering comprehensive solutions for crop planning, monitoring, and analysis.
- To address the challenges and opportunities within the landscape of modern agriculture, it will be essential for agricultural educational and research institutes around the globe to consider the following four dimensions (shown in Fig. 7.2) of agriculture as core educational and research themes for taught and research degree programs related to advanced digital agriculture (https://digitalag.illinois.edu/):

  - *Crops and Animals*: this research theme encompasses a broad range of digital technologies and techniques essential to improve crop/livestock yield/production and environmental sustainability.
  - *Data*: this theme deals with the collection, transfer, storage, processing, analysis, security, privacy, etc. of agricultural data related to crops and animals.
  - *Automation*: this theme revolves around the use of digital technologies that deal with the reduction of human labor for high agriculture production with improvement in operational accuracy, precision, and scalability.

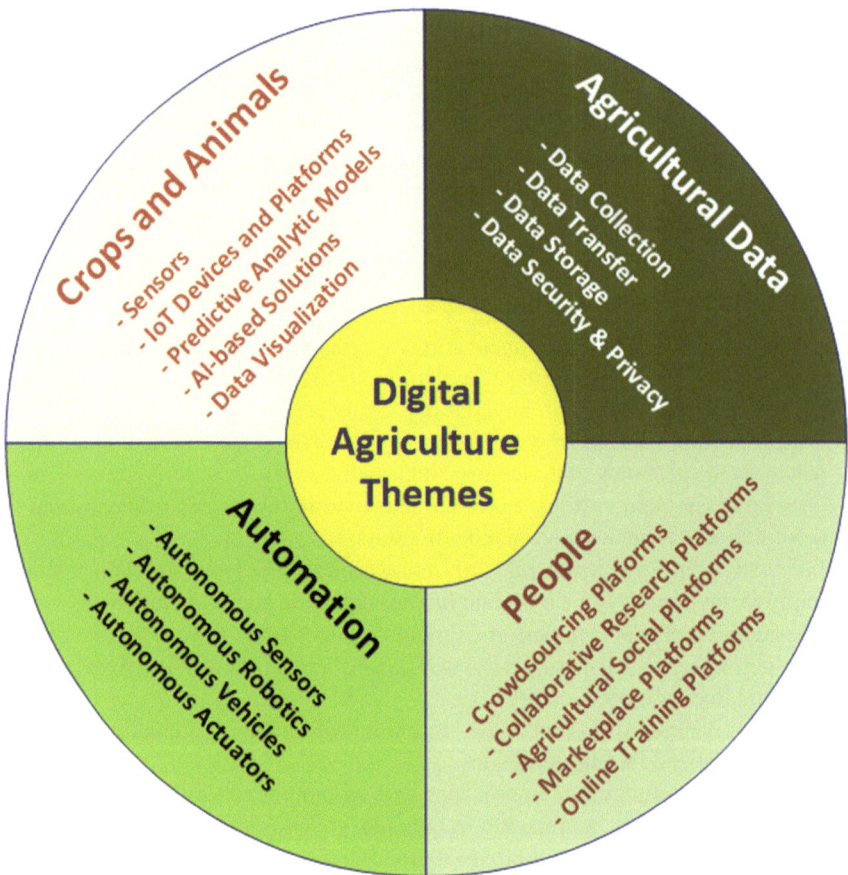

Fig. 7.2 Themes and related digital technologies/techniques/platforms for advanced agriculture degree programs

  – *People in Agriculture*: this theme deals with the use of digital technologies to enable interaction and/or collaboration between scientists, engineers, and agriculturists to improve the social and economic aspects of agricultural products.

Questions

Q7.1: Describe how digital technologies assist agriculturists through the provisioning of personalized strategies for crop management.

Q7.2: What is meant by a closed-cycle agriculture system and explain with the help of an example how digital agriculture supports the implementation of these types of systems.

Q7.3: Elaborate the role of digital technologies in promoting better water management and conservation in agricultural fields.

Q7.4: How do the advancements in digital technologies contribute to the improved monitoring and management of livestock in agricultural fields?

Q7.5: What is meant by enhancement of farmer's agility through the use of digital tools and technologies? Describe with examples.

References

Ahmed R (2013) The web book-build static and dynamic websites. CreateSpace Independent Publishing Platform

Akankasha SM, Vikas D (2014) Expert systems in agriculture: an overview. Int J Sci Technol Eng 1(5):45–49

Alonso G et al (2004) Web services. Web services. Springer, pp 123–149

Bochtis DD, Sørensen CG, Green O (2012) A DSS for planning of soil-sensitive field operations. Decis Support Syst 53(1):66–75

Center for Digital Agriculture at Illinois. https://digitalag.illinois.edu/

Colezea M et al (2018) CLUeFARM: integrated web-service platform for smart farms. Comput Electron Agric 154:134–154

Conesa-Muñoz J et al (2016) A multi-robot sense-act approach to lead to a proper acting in environmental incidents. Sensors 16(8):1269

Dix A et al (2000) Human-computer interaction. Harlow UA

Elijah O et al (2018) An overview of Internet of Things (IoT) and data analytics in agriculture: benefits and challenges. IEEE Internet Things J 5(5):3758–3773

Fikar C (2018) A decision support system to investigate food losses in e-grocery deliveries. Comput Ind Eng 117:282–290

Finlay PN (1990) Decision support systems and expert systems: a comparison of their components and design methodologies. Comput Oper Res 17(6):535–543

Fisher D et al (2012) Interactions with big data analytics. Interactions 19(3):50–59

Ford FN (1985) Decision support systems and expert systems: a comparison. Inf Manage 8(1):21–26

Giusti E, Marsili-Libelli S (2015) A fuzzy decision support system for irrigation and water conservation in agriculture. Environ Model Softw 63:73–86

Gutiérrez F et al (2019) A review of visualisations in agricultural decision support systems: an HCI perspective. Comput Electron Agric 163:104844

Halili F, Ramadani E (2018) Web services: a comparison of soap and rest services. Mod Appl Sci 12(3):175

Kadiyala M et al (2015) An integrated crop model and GIS decision support system for assisting agronomic decision making under climate change. Sci Total Environ 521:123–134

Krit H, Baudin P (1994) D-CAS: an expert system for aid in the appraisal and treatment of diseases of sugar cane. Agriculture et Developpement (France)

Lemmon H (1990) Comax: an expert system for cotton crop management. Comput Sci Econ Manage 3(2):177–185

Mahmoud M, El-Araby K, Rafea A (1995) Limex: an integrated expert system for lime crop management. IFAC Proc Vol 28(4):337–342

Marques MJR (2017) A mobile approach to farmer-computer interaction

Maurya B, Beg MR, Mukherjee S (2013) Expert system design and architecture for farming sector. In: 2013 IEEE conference on information & communication technologies

Navarro-Hellín H et al (2016) A decision support system for managing irrigation in agriculture. Comput Electron Agric 124:121–131

Nevo A, Amir I (1991) CROPLOT—an expert system for determining the suitability of crops to plots. Agric Syst 37(3):225–241

Nixon R (2012) Learning PHP, MySQL, JavaScript, and CSS: a step-by-step guide to creating dynamic websites. O'Reilly Media, Inc

Papazoglou M (2008) Web services: principles and technology. Pearson Education

Plant R et al (1989) CALEX/peaches, an expert system for the diagnosis of peach and nectarine disorders. HortScience 24(4)

Posadas BB et al (2021) Design and evaluation of a crowdsourcing precision agriculture mobile application for Lambsquarters, mission LQ. Agronomy 11(10):1951

Prasad R, Ranjan KR, Sinha A (2006) AMRAPALIKA: An expert system for the diagnosis of pests, diseases, and disorders in Indian mango. Knowl-Based Syst 19(1):9–21

Recio B, Rubio F, Criado JA (2003) A decision support system for farm planning using AgriSupport II. Decis Support Syst 36(2):189–203

Roach J et al (1985) POMME: a computer-based consultation system for apple orchard management using Prolog. Expert Syst 2(2):56–69

Rose DC et al (2018) Involving stakeholders in agricultural decision support systems: Improving user-centred design. Int J Agric Manage 6(1029-2019-924):80–89

Roy J, Ramanujan A (2001) Understanding web services. IT Professional 3(6):69–73

Saizmaa T, Kim H-C (2008) A holistic understanding of HCI perspectives on smart home. In: 2008 Fourth international conference on networked computing and advanced information management

Salah A et al (1993) CITEX: an expert system for citrus crop management. In: Proceedings of the second national expert systems and development workshop (ESADW-93). 1993. Ministry of Agriculture and Land Reclamation Cairo, Egypt

Saunders M et al (1987) GRAPES: an expert system for viticulture in Pennsylvania. In: AI applications in natural resource management (USA)

Soysal M, Bloemhof-Ruwaard JM, Van Der Vorst JG (2014) Modelling food logistics networks with emission considerations: the case of an international beef supply chain. Int J Prod Econ 152:57–70

Sprenkel RK, Momol MT (2002) Distance diagnostic and identification system (DDIS): a new tool for Florida extension diagnostics

Srinivasan R, Engel BA, Paudyal G (1991) Expert system for irrigation management (ESIM). Agric Syst 36(3):297–314

Sriram N, Philip H (2016) Expert system for decision support in agriculture. TNAU Agritech

Talari G et al (2022) State of the art review of big data and web-based decision support systems (DSS) for food safety risk assessment with respect to climate change. Trends Food Sci Technol 126:192–204

Ting S et al (2014) Mining logistics data to assure the quality in a sustainable food supply chain: a case in the red wine industry. Int J Prod Econ 152:200–209

Wenkel K-O et al (2013) LandCaRe DSS–An interactive decision support system for climate change impact assessment and the analysis of potential agricultural land use adaptation strategies. J Environ Manage 127:S168–S183

Yialouris C et al (1997) VEGES—A multilingual expert system for the diagnosis of pests, diseases and nutritional disorders of six greenhouse vegetables. Comput Electron Agric 19(1):55–67

Zhai Z et al (2020) Decision support systems for agriculture 4.0: survey and challenges. Comput Electron Agric 170:105256

Appendix
Answers to Questions Given at the End of Each Chapter

Chapter 1

Q1.1: Yes, like traditional agriculture, digital agriculture can be considered both as science and art. The science of digital agriculture provides a robust basis for developing sustainable agricultural practices through technological innovations and data-driven insights. On the other hand, the art of digital agriculture involves the adaptation of digital technologies to the specific conditions of agricultural fields, climate change, and market demands that ultimately depends on human judgment in terms of intuition, experience, and creativity. For example, the science aspect of precision viticulture in a vineyard includes

- the implementation of soil sensors (to collect soil nutrient content, soil pH, soil moisture, and temperature),
- the deployment of IoT devices (i.e., weather stations and drones/satellites) to collect real-time data on temperature, humidity, wind speed, rainfall, and early signs of disease or pest attacks.
- the application of suitable machine/deep learning algorithms to predict optimal harvest time, pest outbreak, and disease spread.
- the automation of irrigation systems to adjust water delivery in real-time.

On the other hand, the art aspect of precision viticulture in a vineyard involves the intuition and experience of agriculturists related to the adoption of the appropriate sensors, IoT devices, and data analytic technologies while considering the diverse and dynamic environments (i.e., size, location, topography, climate, and market demand) around the specific vineyard.

Q1.2: Precision farming and smart farming (often used interchangeably) have distinct differences. Precision farming predominantly focuses on optimization of field-level management or efficient use of farm resources i.e., water, fertilizer, pesticides, farm machinery, etc. through the use of digital devices, techniques and technologies i.e., sensors, data analytics, GPS technology, etc. Therefore, the main goal of precision

M. A. Iqbal, *Digital Agriculture*, SpringerBriefs in Agriculture, https://doi.org/10.1007/978-3-031-67679-6

farming is to maximize productivity and minimize waste through precise, site-specific management. On the other hand, smart farming encompasses a broader scope, integrating advanced technologies such as IoT, AI, robotics, and big data across the entire agricultural ecosystem. Therefore, smart farming not only improves field-level precision but emphasizes on the creation of interconnected and intelligent agricultural system by connecting various aspects of farm management i.e., automated farm monitoring, farm machinery automation, livestock monitoring, and supply chain optimization.

Q1.3: Along with the rapid population growth and scarcity of agricultural land, the evolution of Digital Agriculture is propelled by several other crucial driving factors i.e., demand for remote monitoring/management (automation), higher productivity (or consumer demand), resource optimization, cost reduction, and minimization of environmental effects.

Q1.4: Both revolutions i.e., Agriculture 4.0 and Agriculture 5.0 mainly emphasize the development of ICT-based (farm management) agricultural systems. However, the focus of Agriculture 4.0 is to automate the processes of precision agriculture and smart farming through the use of robotics, BigData, and IoT technologies. On the other hand, Agriculture 5.0 based on the foundations of Agriculture 4.0 represents the next leap in agricultural innovations. Focusing on human-centric and sustainable approaches, Agriculture 5.0 implies the utilization of cutting-edge techniques and technologies (i.e., advanced AI algorithms, Blockchain, virtual reality, personalization, etc.). Therefore, Agriculture 5.0 represents a holistic approach to farming that put stronger emphasis on sustainable agricultural practices with a clear aim to enhance agricultural productivity and efficient utilization of farm resources.

Q1.5: By enabling more efficient, precise, and sustainable agricultural operations, digital technologies promote environmentally friendly farming practices. For example,

- sensors (soil and environmental) ensure the efficient and real-time utilization of farm resources i.e., the right amount of fertilizer and pesticide application in the agriculture field. The efficient use of farm inputs leads to lower emissions and decreased contamination of natural ecosystems.
- GPS, GIS, Drones, and Satellite imagery providing real-time data on spatial variability, soil conditions, crop health, and pest-infestation areas allow farmers to use resources (water, fertilizer, pesticide) only when and where they are needed and ultimately enhance soil fertility and reduce soil erosion and degradation.
- Predictive analytics and data-driven insights help agriculturists to anticipate and mitigate issues of adverse weather conditions and pest infestation through the adoption of best farming practices that promote sustainability and productivity.
- The use of robots reduces soil compaction, the use of chemical herbicides, and human interventions that ultimately lead to more consistent and efficient farming practices.

- Enhanced traceability of agricultural products through the use of Blockchain helps in reducing food spoilage and waste. Moreover, blockchain technology ensures that farming practices meet environmental standards, and informed consumers support sustainable farming practices by making eco-friendly choices.
- The use of renewable energy sources lowers greenhouse gas emissions.

Chapter 2

Q2.1: Agricultural Stakeholders include farmers, ranchers, seed providers, pesticides and fertilizer companies, agricultural consultants, retailers, advisors, agricultural extension workers, aggregators, distributors, marketers, transporters, animal feed companies, veterinarians, agricultural credit and financial institutions, etc. The digitalization of agricultural operations improves profitability, efficiency, and sustainability across the agricultural value chain and provides benefits to all involved stakeholders. Below are the details of how agricultural stakeholders benefit from the digitalization of agricultural operations.

- The accessibility of real-time data on crop health, soil conditions, and weather patterns (using digital technologies) helps farmers and ranchers gain higher yields and production through the precise management of farm resources, reducing waste. Moreover, robotics and automation of farm machinery improve profitability, reduce labor costs, and sustainable farming practices.
- Advanced data analytics platforms help seed providers by enabling the creation of new seed varieties that are more resilient to climate change, pests, and diseases.
- Pesticide and fertilizer companies utilize digital technologies for the development of more targeted products.
- Digital tools and technologies assist agricultural consultants and advisors in providing more accurate and timely guidance to farmers.
- Blockchain technology helps retailers with improved supply chain transparency and ensures cost savings and higher customer satisfaction.
- Real-time tracking of products improves supply chain efficiency that ultimately enables agricultural distributors and marketers to optimize logistics. Moreover, the identification of market trends and consumer preferences by data analytics helps distributors to develop better digital marketing strategies by offering more effective promotions to increase sales.
- Real-time tracking and route optimization offered by digital platforms help transporters with improved delivery and reduced transportation costs which ultimately ensure that agricultural products are transported efficiently in better condition with minimum spoilage and loss.
- Data-driven insights help animal feed companies and veterinarians to develop better feed products and health management practices (i.e., remote health monitoring and timely cost-effective care).
- Data-driven insights enable agricultural credit and financial institutions to offer more effective financial support (in terms of reduced lending risks).

Q2.2: Consider a smart farming dairy system as an IoT-based Digital Agriculture Ecosystem where microphones and video cameras have been placed strategically in the barn and associated grazing areas to monitor the health and behavior of the cows. The sounds captured by the microphones are continuously analyzed by an AI-enabled system which is able to detect the signs of animal distress, pain, stress, or respiratory issues by detecting changes in cattle's vocalizations and abnormal breathing sounds. Also, microphones help ranchers to detect predator intrusions. Similarly, the footage captured by video cameras is continuously analyzed by an AI-enabled system to detect early signs of illness by observing unusual behavior patterns i.e., changes in cow's posture and movement in the grazing field. Moreover, video cameras also help ranchers with remote monitoring of the cows' feeding patterns which are critical to maintaining the quantity and quality of milk production.

Q2.3: A variety of digital technologies and techniques are involved in effectively detecting plant diseases and pest attacks in agricultural fields. These technologies assist agriculturists in providing comprehensive monitoring and early warning systems.

Sensors

Deployment of various sensors to monitor environmental conditions (i.e., soil moisture, temperature, humidity, etc.) helpful in the identification of favorable conditions for disease outbreaks and pest attacks. For example, changes in humidity levels (detected by humidity sensors) in certain areas of agricultural fields can signal a warning about the rapid growth of specific varieties of Fungi in those areas.

Imaging Technologies

Advanced imaging technologies i.e., multispectral and hyperspectral imaging along with the availability of high-resolution cameras can help agriculturists to detect early signs of diseases in a crop.

Artificial Intelligence and Data Analytics

The extraordinary predictive capabilities of advanced machine/deep learning algorithms can detect anomalies (indicating specific disease outbreaks or pest attacks) in collected data and alert farmers to take preventive or corrective actions timely.

Remote Sensing

Equipped with advanced high-resolution cameras, remote sensing technologies i.e., drones and satellites help to capture images in agricultural fields that can ultimately analyzed to detect early signs of plant stress and diseases.

Geographical Information Systems (GIS)

GIS technology is designed to present spatial or geographical data. The overlaying of sensor data and imaging results on a digital map helps agriculturists to pinpoint disease-affected areas of agricultural fields.

Cloud Computing

The ubiquitous availability of large volumes of collected agriculture data and the high computational power of cloud servers enable agriculturists to access and analyze data for crop disease detection remotely from anywhere in approximately real-time.

Smart Phones and Mobile Applications

Pervasive smartphone availability with plant disease detection and identification apps allows farmers not only to capture and transfer field data for disease analysis but also to receive notifications about potential threats.

Q2.4: (a)The architecture diagram of this IoT-based system is shown below.

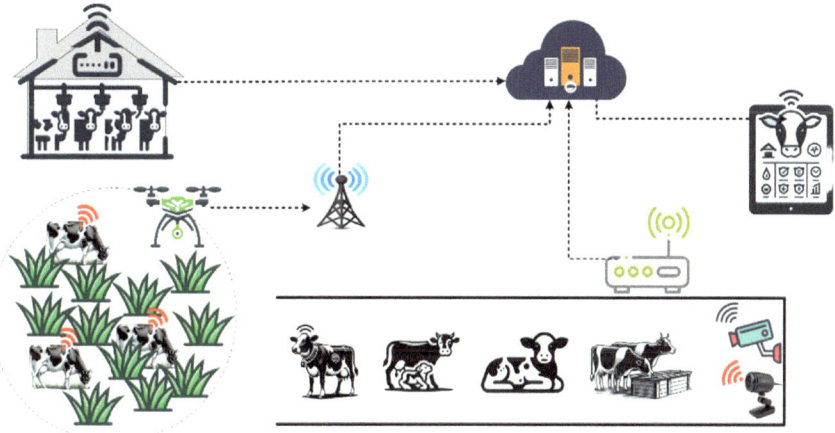

(b) The labeled architecture diagram of this IoT-based system is shown below.

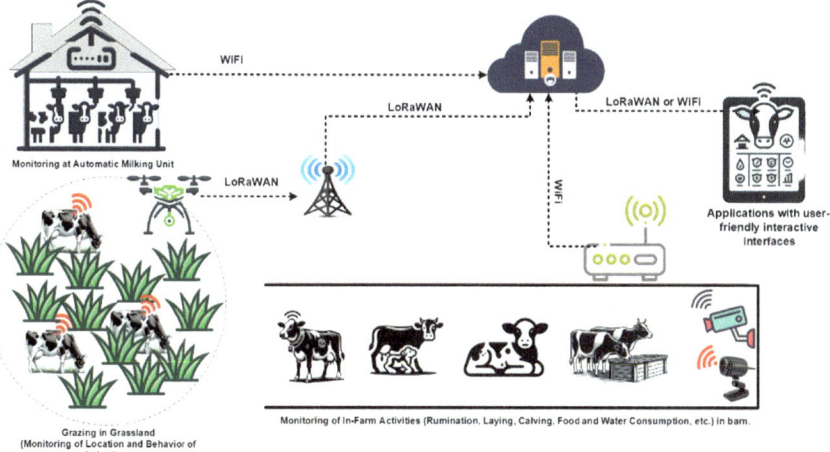

(c) Various types of sensors and digital devices required for this IoT-based dairy farm monitoring system are

Wearable Sensors on Cow(s)

Temperature sensors, Heart rate sensors, GPS trackers, activity sensors (e.g., pedometers, accelerometers, etc.)

Environmental Sensors in Barn and Pasture

Temperature and Humidity Sensors, Light Sensors, Gas Sensors (e.g., ammonia detection).

Data Collection Gateways

Local gateways to collect sensor data via different types of wireless communication (e.g., Bluetooth, ZigBee, WiFi, etc.)

Cloud Servers

To store large volumes of gateways' transmitted data for analysis.

Data Analytics Techniques

AI, Machine learning, and deep learning algorithms to process data for the optimization of feeding schedules and prediction of heath issues that ultimately improve quantity and quality of milk production.

Control Systems

Automated control systems for environmental, feeding, and milking adjustments.

End-User Devices and Interfaces

Smartphones and desktop devices with user-friendly interface mobile applications and web dashboards to monitor and get real-time notifications and results.

Q2.5: The realization of a comprehensive monitoring system in Tomato and Carrot Field depends on a variety of sensors, digital devices, and technologies. Names and brief descriptions of available options for required digital components have been discussed in the following table.

Digital m	Type	Example	Description
Soil moisture sensors	IoT-Based Soil Moisture Sensors	Libelium Waspmote Smart Agriculture Soil Moisture Sensor	Useful for large-scale smart farms
Soil nutrients sensors	Spectroscopy-Based Sensors	SoilOptix	These sensors use gamma-ray spectroscopy to measure multiple nutrients and soil properties

(continued)

(continued)

Digital m	Type	Example	Description
Temperature sensors	Thermocouples	Type T Thermocouple	Measure soil and air temperatures in various smart farm settings
	Wireless Temperature Sensors	HOBO U23 Pro v2 Temperature	Provide real-time temperature data these sensors can be integrated with other environmental sensors
	Soil Temperature Sensors	Decagon 5TE Soil Moisture, Temperature, and EC Sensor	Measure soil temperature along with soil moisture
Humidity sensors	Capacitive Humidity Sensors	Sensirion SHT31	Measure humidity and temperature
	Digital Humidity Sensors	DHT22 (AM2302)	Measure both humidity and temperature and are commonly used in small-scale smart farming setups and are easy to interface with microcontrollers
	Wireless Humidity Sensors	Libelium Waspmote Smart Agriculture Sensor	Provide real-time humidity data wirelessly and are good for large-scale farm monitoring
Rainfall sensors	Tipping Bucket Rain Gauges	Davis Instruments 7852 Rain Collector	Measures rainfall by capturing water in a bucket
	Wireless Rain Gauges	AcuRite 01089 M Rain Gauge with Wireless Display	Measures rainfall and transmits data wirelessly to a remote display in small-scale farms
	IoT-Based Rain Gauges	Libelium Waspmote Smart Agriculture Sensor with Rain Gauge	Provide real-time rainfall data in large-scale farming setups and integrated with cloud-based analytics platforms
Windspeed sensors	IoT-Based Anemometers	Libelium Waspmote Smart Agriculture Sensor with Anemometer	Provide real-time wind speed data for large-scale smart farms integrated with cloud-based analytics platforms

(continued)

(continued)

Digital m	Type	Example	Description
Wind direction sensors	IoT-Based Wind Sensors	Libelium Waspmote Smart Agriculture Sensor with Wind Vane	Integrated with cloud-based analytics platforms use to collect real-time wind direction data in large-scale farms
Weather station	Stationary Agricultural Weather Stations	Spectrum Tech. WatchDog Weather Station	Comprehensive device setups that is used to measure temperature, humidity, rainfall, wind speed, wind direction, etc., and sometimes can be integrated with irrigation systems
		RainWise AgroMET Weather Station	
		Davis Vantage Pro2	
		Ambient Weather WS-2902	
Communication	Long-range Communication Technologies	LoRaWAN	Low-power and long-range communication allows sensors and IoT devices to communicate over long distances
Cloud	Centralized Cloud Setup	Microsoft Azure FarmBeats	A cloud platform that leverages IoT and AI to provide actionable insights for farmers by integrating data from sensors, drones, and satellites

Chapter 3

Q3.1: The concept shown in Fig. 3.7 involves the use of drones as remote sensing platforms equipped with RFID-based sensors to collect data from agricultural fields. This type of system's components, working, and benefits have been discussed below.

System components and setup
Drones are equipped with RFID readers to collect data from field sensors
Small wireless RFID sensors are embedded and attached in soil and plants respectively to monitor soil nutrient levels and plant health

System's working
RFID sensors embedded in soil and attached to plants collect and store data about nutrient levels in soil and plant health indicators. The drone flies over the agricultural fields at pre-defined time intervals on a pre-programmed path to collect stored data wirelessly from soil and plant sensors. The data collected by drones is then transmitted to cloud for analysis using different wireless communication technologies i.e., Bluetooth, WiFi, etc.

(continued)

(continued)

Benefits
The advantages of these types of setups include
- Efficient (faster) data collection from agricultural fields including the places hard-to-reach
- Remote monitoring of field conditions to make timely decisions
- Better resource management by precise identification of areas where farm resources are required to apply
- Minimize the chances of human errors along with reduced labor costs

Q3.2: The types of data and corresponding sensors for IoT-based digital aquaponic systems have been elaborated below.

Fish tank data and corresponding sensors
Water pH Levels: Appropriate pH level is crucial for both fish and plants
Sensors: pH sensors or pH probes
Water Temperature: Appropriate temperature ranges for both fish and plants
Sensors: Thermocouples or water temperature sensors
Dissolved Oxygen Level: Adequate Oxygen level is critical for fish life and the overall system's balance
Sensors: Dissolved oxygen sensors or optical DO sensors
Water Turbidity: indicate the presence of suspended solids or algae
Sensors: Turbidity sensors
Ammonia/Nitrate Levels: High levels of Ammonia, Nitrite, and Nitrate adversely affect fish and plant health
Sensors: ammonia, nitrite, and nitrate sensors
Fish Behavior Data: monitoring of fish movements and feeding patterns ensures good health of fish
Sensors: Underwater cameras, motion sensors, automatic feeders with built-in sensors

Grow bed data and corresponding sensors
Light Intensity: adequate light required for proper photosynthesis
Sensors: Light sensors or PAR (Photosynthetically Active Radiation) sensors
Temperature and Humidity: Air temperature and humidity are critical for plant growth
Sensors: Temperature and humidity sensors
CO_2 levels: CO_2 concentration in air is critical for plant photosynthesis
Sensors: CO_2 sensors

Water level and flow sensors
Maintaining appropriate water levels with proper circulation between the aquarium and grow beds is important to ensure a balanced nutrient supply and optimal oxygenation
Sensors: ultrasonic flow sensors and ultrasonic level sensors, flow meters, and float switches

Q3.3: The Figure depicts a comprehensive architecture of a Digital Dairy Farming System integrating various digital technologies. The use of these technologies along with the flow of data from IoT sensors and digital devices to the cloud has been explained below.

Usage of Sensors and Digital Devices for Data Acquisition

Wearable Sensors on Cows:

- Temperature Sensors to monitor cow's body temperature to detect signs of illness

- Heart Rate Sensors to assess overall health and stress levels in cows
- GPS Trackers to locate cows in grasslands and monitor their grazing patterns
- Activity Sensors to track cows' movement and behavior for health monitoring perspective.

Environmental Sensors

- Temperature and Humidity Sensors monitor barn/pasture ambient conditions to ensure an appropriate environment for cows
- Gas Sensors to detect harmful gas levels in the barn to maintain air quality

Monitoring and Data Collection Devices

- Drones for aerial monitoring of overall farm conditions and cows' movement in the farmland
- Weather Stations to collect real-time data on weather conditions, helping in making decisions related to feeding and milking schedules.

Automated Feeding and Milking Systems

- Feeding Systems to automate feeding schedules based on cows' needs and their health
- Milking Machines (equipped with sensors) to monitor milk quality, volume, etc.

Energy Management Systems: used to monitor energy usage within the farm.

Data Flow in IoT-based Digital Dairy Farming System

Data Acquisition and Transmission: In this phase, data is collected from wearable cow sensors, environmental sensors, drones, and weather stations to measure and monitor animal health, movement, behavior, environmental conditions, etc., and transmitted (through various types of communication technologies i.e., Bluetooth, ZigBee, WiFi, etc.) to local gateways.

Data Aggregation and Transmission at Local Gateways: In this phase, data from various sensors is aggregated at local gateways and transmitted to the cloud using different wireless technologies i.e., 4G/5G, LoRaWAN, etc. for storage and analysis.

Data Storage and Analysis: In this phase, heterogeneous related data in large amounts are stored for analysis to detect anomalies, predict trends, generate actionable insights, etc.

Data Access and Automated Adjustments: In this phase, ranchers can access processed data, real-time alerts/notifications on their mobile applications and web dashboards. Based on processed or analyzed data, automated systems make necessary adjustments i.e., feed scheduling.

Q3.4: Considering two real-life scenarios, the use of video cameras and microphones to monitor livestock activities has been explained below with the help of two examples.

Microphone Usage: Automatic Identification of Coughing Animal

Cough is one of the central elements that is considered an early symptom for the diagnosis of various diseases especially pneumonia and other respiratory diseases, which are the main causes of animal mortality and low farm productivity. Through audio analysis, it becomes possible to distinguish cough from other vocal manifestations and ultimately help to interpret the health conditions of animals/birds in a barn/ flock. Different acoustic parameters i.e., peak coughing sound frequency, coughing time, coughing duration, the time interval between two coughing occurrences, etc. have been considered to discriminate infectious cough from non-infectious cough.

Video Camera Usage: Weight Estimation Using Vision Tools

Automatic identification of low-weight animals by measuring the sizes of moving animals is possible through the analysis of camera videos. In these techniques, it is required to detect areas greater than the size of an individual normal animal (even in a group of animals close and touching each other) by using HSV (Hue, Saturation, Value) color information. If the size of an animal is found abnormal than the experimentally calculated threshold value, then the animal is marked as a low-weight animal.

Q3.5: To monitor low temperature, high humidity, low luminosity, and decreased oxygen levels in the soil that hinder the growth of plantain development, various types of sensors and IoT devices can be implemented as discussed below:

Environmental Monitoring Sensors and Devices

Temperature sensors
Device: digital thermometer
Use: measure ambient temperature in the plantain field

Humidity sensors
Device: hygrometer, capacitive humidity sensor
Use: monitor the humidity levels to prevent high humidity that hinders plantain development

Light sensors (luminosity sensors)
Device: Photometer, Light Dependent Resistor (LDR)
Use: Measure the intensity of light and ensure sufficient luminosity for photosynthesis

Weather stations
Device: Integrated weather station
Use: Collect comprehensive data (i.e., temperature, humidity, wind speed, and solar radiation, etc.) to prevent adverse effects of environmental conditions

Soil Monitoring Sensors and Devices

Soil temperature sensors
Device: Soil thermometer, soil temperature probe, RTD (Resistance Temperature Detector)
Use: Monitor soil temperature of the soil to ensure the optimal range for Plantain growth

(continued)

(continued)

Soil moisture sensors Device: Tensiometer Description: Measure soil moisture to prevent overwatering of the Plantain field	

Soil oxygen sensors   Device: Electrochemical oxygen sensor, optical oxygen sensor   Use: Detect soil oxygen level to prevent adverse effects on Plantain root development due to hypoxic conditions

Growth Monitoring Sensors and Devices

Device: LiDAR (Light Detection and Ranging) Sensors   Use: create 3D maps of Plantain field and assist in measuring plant height, canopy volume, etc.
Device: Multispectral and Hyperspectral Cameras   Use: Multispectral and Hyperspectral images help in detecting changes in leaf color and chlorophyll that are indicators of plant growth and health
Device: Dendrometer sensors   Use: measure the growth of Plantain growth by measuring changes in stem diameter

Chapter 4

Q4.1: In general, P2P communications refer to direct communication between two devices. Similarly, in the context of LoRa technology, P2P communication refers to direct communication between two LoRa-enabled devices without any need for intermediate network infrastructure (i.e., gateways). For example, LoRa-enabled soil moisture sensors deployed at various points in a remote agricultural field measure soil moisture and transmit data directly to LoRa-enabled displays/monitoring devices.

Q4.2: The low-duty cycle devices or sensors wake up periodically and spend most of their time in sleep mode. Therefore, the extended battery life of sensors (devices) is the main advantage as these devices reduce power consumption significantly. One of the examples from an agricultural scenario is the realization of an irrigation management system with low-duty ZigBee-enabled soil moisture sensors deployment in a Vineyard. This type of setup ensures the extended battery life of ZigBee sensors in large-scale agricultural fields.

Q4.3: In a small-scale Tomato field, a combination of Bluetooth, ZigBee, and Wi-Fi is ideal for short-range applications to optimize monitoring and management practices. However, for a large-scale Tomato field, LoRa, NB-IoT, and Sigfox are excellent for wide-area, low-power sensor networks. The integration of these technologies helps farmers to improve efficiency, optimize resource usage, enhance productivity, and improve overall management of Tomato fields.

Q4.4: Maize (or Corn) crop is often grown in large-scale agricultural fields covering several hectares. Therefore, LoRaWAN is highly recommended for these types of large-scale IoT-based agriculture environments because

- LoRaWAN supports data communication over several Km covering the entire field without the need for extensive infrastructure.
- LoRaWAN-enabled sensors deployed in Maize field can operate for longer periods of time without frequent sensor or battery replacements.
- LoRaWAN networks are highly scalable and therefore it would not be difficult to support the expansion of maize crop systems.
- LoRaWAN provides robust and reliable communication in (hash) challenging agricultural environments.
- LoRaWAN is compatible with a wide range of sensors (i.e., temperature, humidity, soil moisture, etc.) and IoT devices.

Q4.5: Reasons for selecting various technologies for this particular scenario have been discussed below.

Rationale for using WiFi

WiFi facilitates remote monitoring by providing a high data rate that is required for video monitoring and transferring high-resolution sensor data. Moreover, it is convenient and cost-effective as most nurseries have existing WiFi networks.

Rationale for using ZigBee

The short-range low-power consumption ZigBee is suitable for sensors that require less bandwidth to transfer data from nurseries i.e., soil moisture sensors, soil temperature sensors, humidity sensors, etc. Due to low power consumption, it is appropriate for sensors to be operated in nursery outdoor areas without frequent battery replacements. Moreover, in a large-scale flower nursery setting, ZigBee devices can be used as repeaters to extend coverage for robust communication.

Rationale for using LoRaWAN

The long-range and low-power consumption capabilities of LoRaWAN with minimal infrastructure make it suitable for large nurseries with related outdoor areas to communicate over several kilometers. Moreover, it is also suitable to transmit small data packets containing data collected from pH sensors or weather station devices.

Chapter 5

Q5.1: Yes, the deployment of edge computing (in addition to cloud infrastructure) for the given scenario significantly enhances the responsiveness and efficiency of the system. Edge devices (i.e., smart gateways or local servers equipped with machine learning models in an edge-cloud solution) deployed on the farm perform initial data processing and analysis close to the data source(s). Reduced latency and optimized use of network resources are two main benefits of hybrid edge-cloud solutions. This can be comprehended from an example scenario where sensors on a cow transmit vital signs data to smart gateways that can process received data in real-time to detect anomalies and specific animal behaviors. The edge devices in this scenario are capable of sending instant alerts to the farmer's smartphone in case of anomaly detection. For this type of scenario, the cloud platform analyzes aggregated data to perform predictive analysis to identify long-term trends.

Q5.2: The architecture diagram of the developed mobile application system for recognizing pests using some machine learning techniques implemented in a cloud computing system is given below.

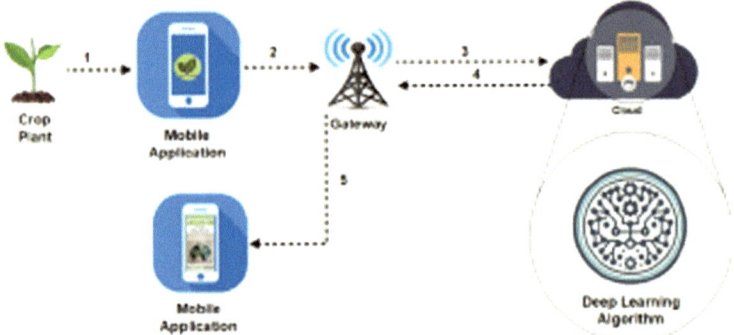

The basic steps for this type of end-to-end system are described below.

1. Using the Android application on the mobile phone, the farmer (using a smartphone) can take pictures of crop plants in the agricultural field.
2. Using appropriate wireless communication technology, the taken photo is transferred to cloud servers via intermediate gateways.
3. Upon receiving the image, servers in the cloud invoke a deep-learning module (to process the image for disease detection).
4. The deep learning module processes the image and sends the result back to the smartphone application using appropriate wireless communication technologies.
5. The application on the smartphone displays the result.

Q5.3:

Scenario 1: Use of Autonomous Farm Vehicles in Agricultural Fields

Fog nodes providing local processing near self-driving tractors and drones in agricultural fields support autonomous operations in real-time. This real-time processing ensures that vehicles can respond quickly to changing conditions.

Scenario 2: Livestock Monitoring in Large-scale Pasture Lands

Fog nodes providing local processing near pasture lands assist ranchers in various ways i.e.,

- Real-time identification of cattle illness or abnormal behavior
- Real-time alerts to farmers or veterinarians for further investigations
- Handling large numbers of cattle without overwhelming central cloud servers

Q5.4: Edge computing significantly improves the efficiency and effectiveness of smart farming applications by

- Reducing communication cost
- Enabling real-time processing
- Enabling context-aware decision-making
- Optimizing the usage of farm resources
- Ensuring system reliability

Q5.5:

Size of the image in bits $(L_1) = 256$ MB $= 256 \times 8 \times 10^6$ bits.
    Size of response data in bits $(L_2) = 32$ MB $= 32 \times 8 \times 10^6$ bits.
    Duplex link data rate for Fog server $(R_f) = 128$ MB/Sec.
    Duplex link data rate for Cloud server $(R_c) = 128$ MB/Sec.
    The generic formula to calculate the overall delay from source to destination and back from destination to the source is given in Eq. 1.

$$
\begin{aligned}
\textit{Overall Delay} \; = \; & \textit{Transmission Delay}\,1\;(T_{d1}) \; + \; \textit{Propagation Delay}\,1\;(P_{d1}) \\
& + \; \textit{Processing Delay}\;(P_{rd}) \; + \; \textit{Transmission Delay}\,2\;(T_{d2}) \\
& + \; \textit{Propagation Delay}\,2\;(P_{d2})
\end{aligned} \tag{1}
$$

The formula to calculate Transmission Delay (T_{d1} or T_{d2}) is given in Eq. 2.

$$
\textit{Transmission Delay}\;(Td) \; = \; \textit{Data Size in bits}\;(L)/\textit{Data rate on the link}\;\text{(bits/sec)} \tag{2}
$$

T_{d1} and T_{d2} in Eq. 1 represent transmission delay from sender (Tractor device) to receiver (Fog or Cloud server) and from the destination (Fog or Cloud server) to source (Tractor device), respectively.
    Processing Delay (P_{rd}) is fixed for a device (but varies from device to device). Below, we have mentioned it as P_{rdf} (to represent the processing delay on the Fog server) or P_{rdc} (to represent the processing delay on the Cloud server). In this example, both Fog and Cloud servers have the same processing delay which is 10 ms.
    The formula to calculate Propagation Delay (Pd1 or Pd2) is given in Eq. 3.

$$
\begin{aligned}
\textit{Propagation Delay}\;(Pd) \; = \; & \textit{Link distance from sender to the receiver in} \\
& \textit{meters}\;(D)/\textit{Speed of link}\;(meters/sec)
\end{aligned} \tag{3}
$$

P_{d1} and P_{d2} in Eq. 1 represent propagation delay from sender (Tractor device) to receiver (Fog or Cloud server) and from receiver (Fog or Cloud server) to sender (Tractor device), respectively. In this example, these delays have already been given in the problem statement.

(a) *Overall delay in the case of Fog computing*

Transmission Delay in the case of sending data from the tractor to the Fog server is

$T_{d1} = L_1/R_f = (256 \times 8 \times 10^6 \text{ bits})/(128 \times 8 \times 10^6 \text{ bits/sec}) = 2 \text{ Sec.}$

Propagation Delay from the tractor to the Fog server $(P_{d1}) = 200 \text{ ms} = 0.2 \text{ Sec}$

This 0.2 Sec P_{d1} is given in the problem statement (that was actually calculated by considering the formula given in Eq. 2).

Processing Delay on Fog Server $(P_{rdf}) = 10 \text{ ms} = 0.01 \text{ Sec}$

Transmission Delay in the case of sending data from the Fog server to the Tractor device is

$T_{d2} = L_2/R_f = (32 \times 8 \times 10^6 \text{ bits})/(128 \times 8 \times 10^6 \text{ bits/sec}) = 0.25 \text{ Sec.}$

Propagation Delay from the Fog server to the Tractor device $(P_{d2}) = 200 \text{ ms} = 0.2 \text{ Sec.}$

This 0.2 Sec P_{d2} is given in the problem statement (that was actually calculated by considering the formula given in Eq. 2).

Overall Delay $= T_{d1} + P_{d1} + P_{rdf} + T_{d2} + P_{d2} = 2 + 0.2 + 0.01 + 0.25 + 0.2 = 2.66 \text{ Sec.}$

(b) *Overall delay in the case of Cloud computing*

Transmission Delay in the case of sending data from the Tractor device to the Cloud server is

$T_{d1} = L_1/R_c = (256 \times 8 \times 10^6 \text{ bits})/(128 \times 8 \times 10^6 \text{ bits/sec}) = 2 \text{ Sec.}$

Propagation Delay from the Tractor device to the Cloud server $(P_{d1}) = 300 \text{ ms} = 0.3 \text{ Sec.}$

This 0.3 Sec P_{d1} is given in the problem statement (that was actually calculated by considering the formula given in Eq. 2).

Processing Delay on Cloud Server $(P_{rdc}) = 5 \text{ ms} = 0.005 \text{ Sec.}$

Transmission Delay in the case of sending data from the Cloud server to the Tractor device is

$T_{d2} = L_2/R_c = (32 \times 8 \times 10^6 \text{ bits})/(128 \times 8 \times 10^6 \text{ bits/sec}) = 0.25 \text{ Sec.}$

Propagation Delay from the Cloud server to the Tractor device $(P_{d2}) = 300 \text{ ms} = 0.3 \text{ Sec.}$

This 0.3 Sec P_{d2} is given in the problem statement (that was actually calculated by considering the formula given in Eq. 2).

$$\text{Overall Delay} = T_{d1} + P_{d1} + P_{rdc} + T_{d2} + P_{d2} = 2 + 0.3 + 0.005 + 0.25 + 0.3 = 2.855 \text{ Sec}$$

Fog computing implementation would be preferred in this case, because the overall delay in the case of Fog computing is less than the overall delay in the case of Cloud computing.

Chapter 6

Q6.1: In agricultural data analytics,

- Primary data sources are collected directly through sensors and IoT devices deployed in agricultural fields and offer specific information about an agricultural entity and surrounding environment
- Secondary data sources derived from primary data sources and offer supplemental data to enhance primary data understanding
- Tertiary data sources provide synthesized overviews and are useful for gaining a broad understanding of agricultural trends/practices in a specific area, province, or country, etc.

The integration of primary, secondary, and tertiary data sources helps agriculturists to enhance farming practices and crop productivity by offering comprehensive insight. The names of these three types of data sources have been mentioned in the table below.

Data source type	Data source name
Primary	Sensors i.e., Soil Moisture Sensors, Soil Nutrient Sensors, Temperature, and Humidity Sensors, etc.
	Drones and Satellites with thermal and hyperspectral cameras
	Weather stations
	Field survey or crop yield data
Secondary	Government and institutional reports i.e., agricultural census data, soil surveys, and maps
	Research publications in academic journals and conference proceedings
	Market data in the form of commodity prices and supply-chain reports
Tertiary	Agricultural encyclopedias and handbooks
	Aggregator databases i.e., World Bank agricultural data, industry reports, and market analysis
	Summarized data from agricultural extension services

Q6.2: Data analytics enhances crop yield and quality by helping farmers with actionable insights that are derived from a huge amount of agricultural BigData collected from various types of sensors and digital devices i.e., temperature/humidity sensors, moisture sensors, drones/satellites with hyperspectral cameras, weather stations, etc. For example, analysis of collected

- soil moisture data helps farmers optimize irrigation schedule
- data from soil nutrient sensors allows farmers to ensure targeted fertilization
- historical and real-time data about pests and disease outbreaks assists farmers in terms of reducing the need for reactive pesticide applications by implementing preventive and integrated pest management strategies

- climate and environmental data allow agriculturists to optimize crop rotation and planting schedules with useful insights to promote sustainable farming practices

Q6.3: Several machine learning algorithms (belonging to categories of supervised, unsupervised, and deep learning) can be implemented in agricultural data analytics to get useful insight. For example,

Support Vector Machines (SVM): Using labeled data training, these supervised machine learning algorithms can classify pest and crop disease data into different categories.

Random Forests: by constructing multiple decision trees, these supervised learning algorithms are able to classify pests and crop diseases.

K-Nearest Neighbors (KNN): considering the class of nearest neighbors, these algorithms can classify crop diseases and pests into different categories.

K-Mean Clustering: by grouping data into clusters this unsupervised learning technique can identify unusual patterns in crop images that may indicate different types of pests and crop diseases.

Convolutional Neural Networks (CNNs): these deep learning algorithms can assist agriculturists in classifying pests and crop diseases using large datasets of labeled images.

Q6.4: Various types of interactive charts allow agriculturists to explore data dynamically. Some preferred types of interactive visualization charts to effectively visualize crop production in different countries across different years have been discussed below.

Multi-Series Line Chart

Lines in these types of charts representing different crops or countries are ideal for identifying trends and patterns in overall crop yield over multiple years as shown in the below figures.

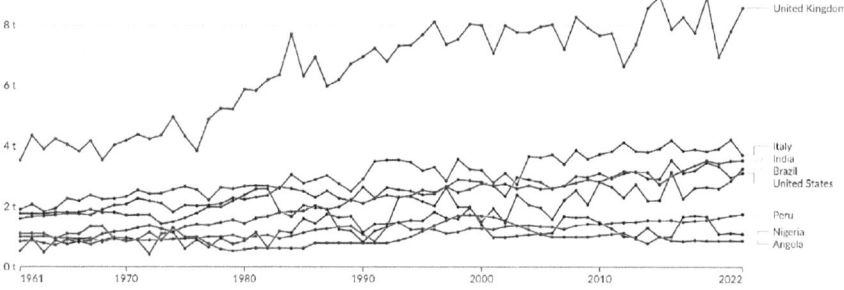

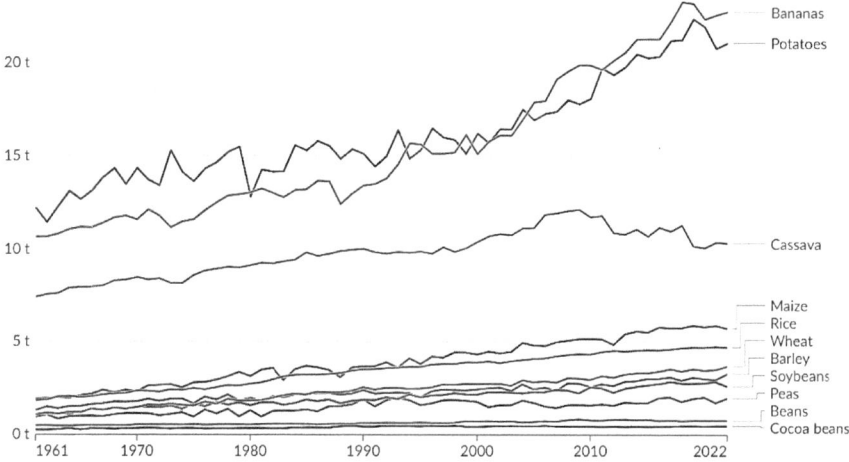

Source Retrieved from: 'https://ourworldindata.org/crop-yields' [Online Resource]

Interactive World Map

With coloring based on the production volume of a specific crop in a specific year, these interactive maps are excellent for comparing regional differences in crop production as shown in the below figure.

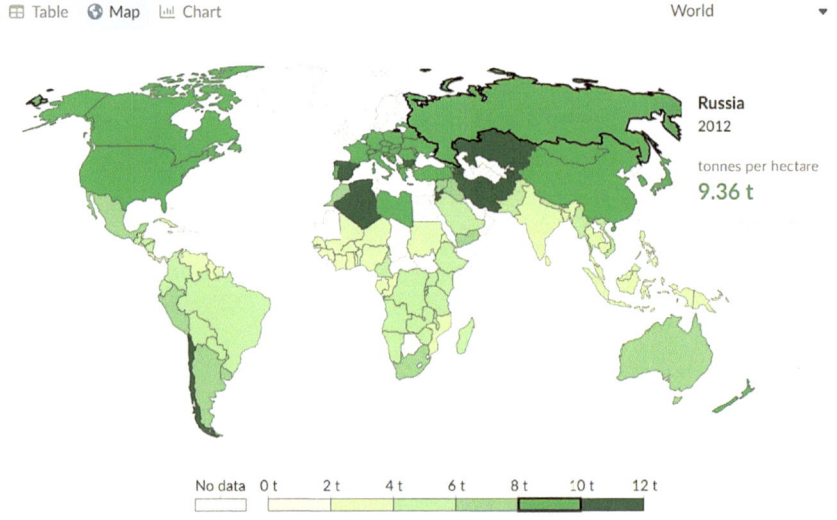

⊞ Table 🌐 Map 📊 Chart   World   ▼

Source Retrieved from: 'https://ourworldindata.org/crop-yields' [Online Resource]

Stacked Area Chart

By displaying crop production levels on top of each other for each country or continent these types of charts are useful for comparing crop production across continents and years as shown below.

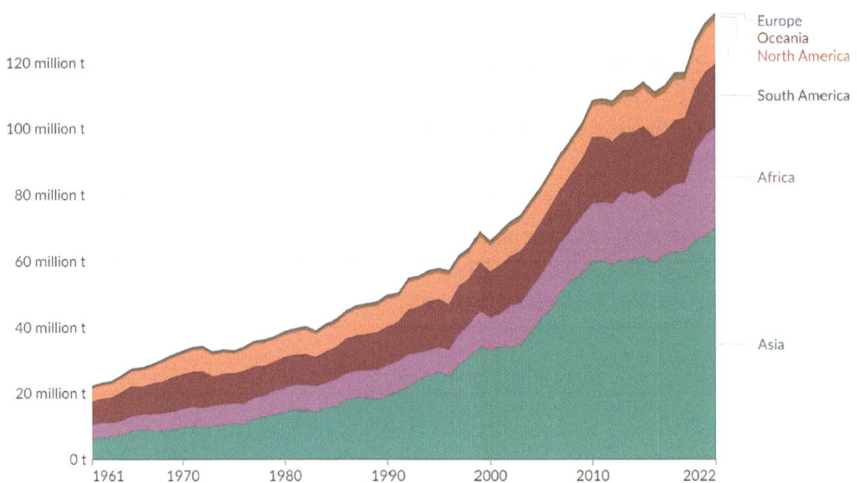

Source Retrieved from: 'https://ourworldindata.org/crop-yields' [Online Resource].

The preferred user-friendly controls to include in interactive charts are Dropdown Menus, Sliders, Tooltips, etc. as highlighted in the below figure.

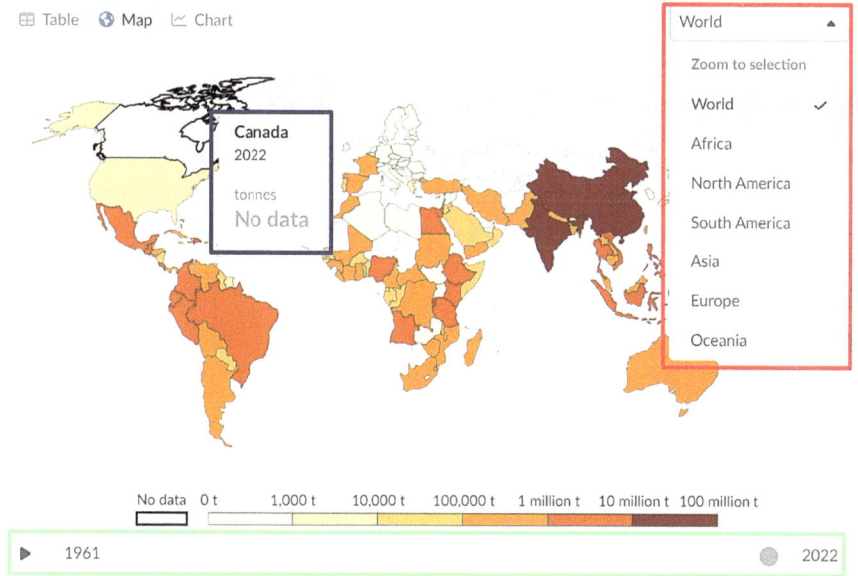

Source Retrieved from: 'https://ourworldindata.org/crop-yields' [Online Resource]

Q6.5: A number of data analytics applications have been developed for livestock management. A few examples are mentioned below.

- *Allflex Livestock Intelligence:* A web application that provides wearable tags, tools, and data analytic platforms to assist farmers by providing real-time identification, monitoring, and tracking of individual animals to manage animal's health, well-being, and productivity.
- *Connecterra (Ida)* is a machine-learning platform that helps ranchers manage herds and optimize animal feeding.
- *CattleEye* is an AI-powered video analytics application that helps ranchers monitor cattle health (via lameness detection and weight changes).
- *Moocall* application by offering tail-mounted sensors supports ranchers in reducing risks associated with birth complications by predicting calving times.
- *HerdDogg* application through the use of smart tags allows ranchers to monitor cattle location and activity.
- *CowManager* application utilizing ear sensors provides actionable insights for farmers to track cattle behavior, feed intake, and health.
- *CowScout* is a health monitoring application that is used by farmers for health monitoring and heat detection in cattle.

Chapter 7

Q7.1: Digital technologies have revolutionized the agriculture industry through the provisioning of personalized strategies for crop management. In this regard, IoT-based agricultural systems (implemented for continuous collection of environmental data) assist agriculturists through receiving personalized recommendations on mobile phones for timely intervention to pest attacks, disease outbreaks, fertilization plans to overcome nutrient deficiencies in soil, adjust irrigation schedules, etc. These personalized alerts for timely interventions and efficient resource utilization ultimately provide profit to farmers in terms of increased crop yield.

Q7.2: A closed-cycle (aka closed-loop or circular) agriculture system refers to a sustainable farming approach where all resources are reused and all nutrients are recycled within the system to minimize the use of external inputs. Therefore, resource recycling, reduced external inputs, and sustainable waste management key characteristics of closed-cycle agriculture systems.

Aquaponics is one of the examples of closed-cycle agriculture that is largely supported by digital agriculture. In Aquaponic systems, fish waste is used as essential plant nutrients and plants filter and clean the water for fish. Therefore, the three main components of aquaponic systems are fish tanks (aquariums), plant-grown beds, and water circulation systems. The use of sensors helps to monitor key system parameters i.e., temperature, humidity, light intensity, pH, ammonia, and nitrate levels in water. On the other hand, based on available sensor data, the use of automated systems helps to regulate water flow, lighting, aeration, temperature, etc. Moreover, the use of data analytics tools helps to provide insights to improve productivity. In this way, precise continuous monitoring, automated control, and real-time data analysis improve the scalability, sustainability, efficiency, and productivity of aquaponic systems.

Q7.3: Digital technologies play an important role in enabling better water management and conservation in agricultural fields. For example,

- Soil moisture sensors reduce water wastage by providing accurate information about when and how much to irrigate a specific piece of land
- Weather sensors help farmers to adjust irrigation schedules by providing information about temperature, humidity, rainfall, etc.
- Large-scale agricultural field monitoring through drones and satellites helps agriculturists with targeted irrigation by identifying areas that require more water.
- Smart irrigation pumps and controllers ensure minimization of over-irrigation.
- Predictive analysis based on historical and real-time data ensure the effective use of watering in agricultural lands.
- Mobile applications assist farmers in getting real-time alerts and making informed decisions to adjust irrigation practices effectively.

Q7.4: Modern livestock farming has become more efficient and effective by enabling real-time monitoring through the use of sensors, automated systems, and actionable insights. For example,

- Wearable sensors (including implantable sensors for body temperature and heart rate, ear tags, collars, etc.) to monitor vital signs that ultimately enables early detection of animal illness and stress.
- RFID and GPS technologies enable ranchers to track the location and movement of livestock ultimately preventing loss.
- GPS data helps manage grazing patterns efficiently by avoiding overgrazing and optimizing pasture use.
- Smart feeders control the right amount of feed to meet the individual needs of animals
- IoT sensors by providing real-time data about animal weight and food consumption ensure precise feeding requirements of individual animals. Precise food consumption lowers the cost and reduces feed wastage.
- Estrus detection sensors timely detect estrus signs in female animals.
- Portable ultrasound devices and imagery help monitor pregnancy and detection of reproductive issues, respectively.
- Climate sensors trigger alerts when environmental parameters fall outside optimal ranges.
- Air quality sensors (by monitoring the level of dangerous gases i.e., ammonia, methane, etc. help to maintain proper environmental conditions that not only prevent animal diseases but also promote animal well-being.
- Data analytic techniques provide insights into herd health and productivity.
- Machine learning algorithms help ranchers to predict animal health issues.

Q7.5: Farmer's agility refers to his ability to quickly

adapt to changes e.g., adjust farming practices such as planting or irrigation schedules through forecasting by the use of digital tools and technologies
make informed decisions e.g., based on predictive analytics quickly decide to apply fertilizer or pest control measures.

Therefore, the adoption of new digital techniques and technologies enables farmers to respond swiftly to changing environmental conditions to optimize farm resource usage and crop production by providing them with automation tools, predictive insights, and real-time data.

Index

© The Editor(s) (if applicable) and The Author(s), under exclusive license   149
to Springer Nature Switzerland AG 2024
M. A. Iqbal, *Digital Agriculture*,
SpringerBriefs in Agriculture, https://doi.org/10.1007/978-3-031-67679-6